# *Foreword*

The Division of Chemical Education (DivCHED) of the American Chemical Society (ACS) established a Committee on Examinations and Tests in 1930. Shortly thereafter, the committee started producing tests for general chemistry. A major goal was to achieve a uniform national standard for demonstrating basic competency in chemistry. Through the intervening years, ACS examinations have become the world standard for evaluating student knowledge in undergraduate chemistry courses, in high school chemistry courses, in course placement for college freshmen, and in establishing chemistry knowledge of entering graduate students.

The Division of Chemical Education reorganized its examinations program in 1987, and founded the ACS DivCHED Examinations Institute, which is responsible for a broad range of materials for chemistry assessment. At the time of this writing, the Examinations Institute is located on the campus of Clemson University.

Since the beginning, each ACS exam has been written by a group of exemplary teachers, all of whom actually teach the course for which the exam is being written. Every item that is accepted for a national ACS exam has been extensively tested, and reviewed by dozens of very capable chemistry teachers.

Not only must every *item* survive careful scrutiny, each entire *test* must be satisfactory. Every national exam must faithfully reflect the most-often-found content of the course it is designed for, and it must be at an appropriate level to challenge the best student while not discouraging the struggling student unnecessarily. The teacher–experts who serve as members of ACS exam committees certify the quality of every ACS exam before it is released.

The care that goes into producing ACS exams may be lost on students who view the exams as foreign and unfamiliar. The purpose of *The Official Guide* is to remove any barriers that might stand in the way of students demonstrating their knowledge of chemistry. The extent to which this goal is achieved will become known only as future generations of chemistry students sit for an ACS exam in general chemistry.

We wish them the best.

*Lucy T. Eubanks*
*I. Dwaine Eubanks*

*Clemson, South Carolina*
*December, 1998*

# Acknowledgements

The unselfish dedication of hundreds of volunteers who contribute their time and expertise make ACS Division of Chemical Education exams and Olympiad exams possible. It is from the reservoir of their work that we have drawn inspiration and examples to produce this book to help students who will be taking an ACS exam. We gratefully acknowledge the efforts of all the past General Chemistry Committee members and Olympiad Task Force members.

*The Official Guide* also benefited from the careful proof reading by several colleagues. We extend our special thanks to these faculty members.

| | |
|---|---|
| Clyde R. Metz | College of Charleston |
| Cheryl Baldwin Frech | The University of Central Oklahoma |
| Mark Freilich | The University of Memphis |
| Dorothy B. Kurland | West Virginia University Institute of Technology |
| David Dever | Macon State College |
| DaFei Feng | San Diego Mesa College |
| M. Elizabeth Derrick | Valdosta State University |
| Diane M. Bunce | The Catholic University of America |
| Jeffrey R. Appling | Clemson University |

The personnel of the ACS Division of Chemical Education Examinations Institute played a central role in helping us to produce *The Official Guide.* A very special thank you for all of the work involved is owed to our staff members.

| | |
|---|---|
| Brenda A. Rathz | Clemson University |
| Sherri P. Morrison | Clemson University |
| W. Sam Burroughs | Clemson University |

While all of these reviewers have been very helpful in finding problems large and small, any remaining errors are solely our responsibility. You can assist us in the preparation of an even better product by notifying the Exams Institute of any errors you may find.

*Lucy T. Eubanks*
*I. Dwaine Eubanks*

*Clemson, South Carolina*
*December, 1998*

# *How to Use This Book*

Every year, students tell us that they really know more than they demonstrate on their final examination in general chemistry. Sometimes the questions are described as "tricky"; sometimes insufficient time is thought to be the problem; and sometimes the material is judged to differ from what was covered in class. These problems most often result from chemistry having been learned as a set of formulas and techniques, rather than as a coherent set of conceptual models that enables comprehension of the submicroscopic world. We urge you, as you are learning chemistry, to strive for genuine understanding of the concepts and models. Simply learning to plug numbers into equations that are meaningless to you is *not* learning chemistry; and that knowledge will not serve you well at exam time.

The major divisions of this book correspond to the common groupings of topics covered by ACS exams for general chemistry. If you are taking an exam that covers the full year of general chemistry, all of the topic groups can be expected on the exam. If you are taking a first-term exam, only the first five topic groups pertain. For a second-term exam, the final five topic groups are most important; although the second-term topics *do* build upon knowledge acquired in the first term.

Each topic group is introduced with a short discussion of the important ideas, concepts, and knowledge that are most frequently stressed in general chemistry courses. *These discussions are not a substitute for studying your textbook, working the problems there, and discussing the challenging ideas with your teachers and fellow students.* Rather, they are reminders of what you have studied and how it fits into a larger understanding of that part of the natural world that we call chemistry.

Next, questions are presented that address those ideas. These questions have been drawn from past ACS exams and U.S. Olympiad exams, and they should give you a good idea of the depth and range of understanding that is expected. Each question is dissected, and you will see how chemists think through each of the questions to reach the intended response. You will also see how choosing various wrong responses reveal misconceptions, careless computation, misapplication of principles, or misunderstandings of the material. Knowing how each incorrect answer is generated will assist you in diagnosing problems with your grasp of the principle being examined.

---

The most effective way to use this book is to answer each **Study Question** before looking at the discussion of the item. Jot down a note of how you arrived at the answer you chose. Next, look at the analysis of the question. Compare your approach with that of the experts. If you missed the item, do you understand why? If you chose the correct response, was it based on understanding or chance? After you have spent time with the **Study Questions**, treat the **Practice Questions** as if they were an actual exam. Allow yourself 50 minutes, and write down your response to each question. Finally, score yourself. Go over the practice questions again. Write down what you needed to know before you could answer the question; and write down how you should think the problem through to reach the intended answer.

---

This book is designed to help you demonstrate your *real* knowledge of chemistry. When you take a general chemistry examination prepared by the American Chemical Society Examinations Institute, you should be permitted to concentrate on demonstrating your knowledge of chemistry, and not on the structure of the examination. We sincerely hope that *The Official Guide* will enrich your study of chemistry, and minimize the trauma of effectively demonstrating what you have learned.

# Sample Instructions

You will find that the front cover of an ACS Exam will have a set of instructions very similar to this. This initial set of instructions is meant for both the faculty member who administers the exam and the student taking the exam. You will be well advised to read the entire set of instructions while waiting for the exam to begin. This sample is from the second semester general chemistry exam released in 1998.

---

**TO THE EXAMINER:** This test is designed to be taken with a special answer sheet on which the student records his or her responses. All answers are to be marked on this answer sheet, not in the test booklet. A pre-punched scoring stencil is available to facilitate grading. Each student should be provided with a test booklet, one answer sheet, and scratch paper; all of which must be turned in at the end of the examination period. The test is to be available to the students **only** during the examination period. For complete instructions refer to the *Directions for Administering Examinations*. Only nonprogrammable calculators are permitted. Norms are based on:

### Score = Number of right answers
### 75 items — 110 minutes

**TO THE STUDENT:** DO NOT WRITE ANYTHING IN THIS BOOKLET! Do not turn this page until your instructor gives the signal to begin. A periodic table and other useful information is on page 2. When you are told to begin work, open the booklet and read the directions on page 3.          **STOCK CODE GC98SR**

---

Note the **restrictions** on the type of **calculators** that you may use and the **time** for administering the exam. These restrictions apply to allow your results to be compared to national norms, ensuring that all students have had the same tools and time to display their knowledge. Your instructor may choose not to follow the calculator or time restriction, particularly if they do not plan to submit your data as part of the national process for calculating norms.

Be sure to notice that scoring is based **only** on the number of right answers. There is no penalty, therefore, for making a reasonable guess even if you are not completely sure of the correct answer. Often you will be able to narrow the choice to two possibilities, improving your odds at success. You will need to keep moving throughout the examination period, for it is to your advantage to attempt every question. Do not assume that the questions become harder as you progress through an ACS Exam. Questions are not grouped by difficulty, but by topic.

You will also be told the **location of the periodic table** and other useful information. In this sample, they are located on page 2 but often these reference materials will be located on the last page or even on a separate sheet called a General Chemistry Data Sheet. In *The Official Guide*, please look on page 113.

Next, here is a sample of the directions you will find at the beginning of an ACS exam.

---

### DIRECTIONS
- When you have selected your answer, blacken the corresponding space on the answer sheet with a soft, black #2 pencil. Make a heavy, full mark, but no stray marks. If you decide to change an answer, erase the unwanted mark very carefully.
- Make no marks in the test booklet. Do all calculations on scratch paper provided by your teacher.
- There is only one correct answer to each question. Any questions for which more than one response has been blackened will not be counted.
- Your score is based solely on the number of questions you answer correctly. It is to your advantage to answer every question.

---

Pay close attention to the mechanical aspects of these directions. Marking your answers without erasures helps to create a very clean answer sheet that can be read without error. As you look at your Scantron® sheet before the end of the exam period, be sure that you check that every question has been attempted, and that only one choice has been made per question. As was the case with the cover instructions, note that your attention is again directed to the fact that the score is based on the total number of questions that you answer correctly. You also can expect a reasonable distribution of **A**, **B**, **C**, and **D** responses, something that is not necessarily true for the distribution of questions in *The Official Guide*.

# Table of Contents

# Atomic Structure

The section of the general chemistry course that deals with atomic structure is designed to describe in qualitative terms the current understanding of the fundamental structure of all matter. Here, atoms are introduced as the basic building blocks of matter. The composition of the atom is described. Frequently, the fundamental experiments that led to the discovery of electrons, protons, and neutrons are laid out in some detail.

The development of quantum mechanics enabled chemists to describe electron energies and locations outside the nucleus more accurately than was possible with the planetary model for the atom. The meanings and implications of quantum numbers, photons, electromagnetic radiation, and radial probability distributions are central to describing the atom in terms of quantum mechanics. Other central ideas include the *aufbau* principle and the uncertainty principle.

Symbolic representations of atomic structure and three-dimensional representations of electron probability regions enable chemists to deal with the behavior of atoms in terms of current atomic theory and to communicate chemical information more clearly.

## Study Questions

**AS-1.** An atom of strontium–90, $^{90}_{38}\text{Sr}$, contains

- **(A)** 38 electrons, 38 protons, 52 neutrons.
- **(B)** 38 electrons, 38 protons, 90 neutrons.
- **(C)** 52 electrons, 52 protons, 38 neutrons.
- **(D)** 52 electrons, 38 protons, 38 neutrons.

*Knowledge Required:* (1) The symbolic representation of isotopes, with the atomic number of an element written as a left-hand subscript, the mass number written as a left-hand superscript, and the ionic charge (if any) written as a right-hand superscript. (2) The fact that the mass number is the sum of the number of protons and neutrons.

*Thinking it Through:* The atomic number of 38 indicates 38 protons, eliminating choice (**C**). This is a neutral atom, so there are also 38 electrons, eliminating choice (**D**). The mass number of 90 for strontium eliminates choice (**B**), which corresponds to a mass number of 128. The correct number of neutrons is found by subtracting the atomic number (38) from the mass number (90) giving a value of 52, which is found in choice (**A**). *Note:* There is redundant information in this question; the term "strontium–90" gives the same information as the symbolic representation, $^{90}_{38}\text{Sr}$ if you know the symbol for strontium and can find its atomic number on the periodic table.

**AS-2.** What do these have in common?

$$^{20}\text{Ne} \qquad ^{19}\text{F}^- \qquad ^{24}\text{Mg}^{2+}$$

- **(A)** the same number of protons
- **(B)** the same number of neutrons
- **(C)** the same number of electrons
- **(D)** the same size

***Knowledge Required:*** (1) The symbolic representation of isotopes, with the atomic number of an element written as a left-hand subscript, the mass number written as a left-hand superscript, and the ionic charge (if any) written as a right-hand superscript. (2) The fact that the mass number is the sum of the number of protons plus neutrons.

***Thinking it Through:*** The periodic table contains the atomic number of each element, which indicates both the number of protons in the nucleus and the number of electrons *for a neutral atom*. Negative ions have one or more additional electrons, and positive ions are short by one or more electrons. Also, subtracting the atomic number from the mass number gives the number of neutrons. These three species have different atomic numbers (10, 9, and 12), so the number of protons cannot be the same, eliminating choice (**A**). Subtracting each atomic number from the given mass number (20, 19, and 24) gives different numbers of neutrons in each case (10, 10, and 12), eliminating choice (**B**). Considering the number of electrons, the neutral atom neon has 10; fluorine has 9 in a neutral atom—and 10 in a fluoride ion; magnesium has 12 in a neutral atom—and 10 in a +2 magnesium ion. The number of electrons is thus the same for all three. Such species are called *isoelectronic*. Note that $Na^+$, $O^{2-}$, and $Al^{3+}$ are also isoelectronic with the given species. Choice (**D**) cannot be true; because, given equal numbers of electrons, those nuclei with greater positive charge will attract electrons more strongly than those with less positive charge.

---

**AS-3.**    Which pair represents isotopes?

(A)    $^{54}_{24}Cr$ and $^{54}_{26}Fe$

(B)    $^{235}_{92}U$ and $^{238}_{92}U$

(C)    $^{116}_{48}Cd$ and $^{116}_{50}Sn$

(D)    $^{239}_{93}Np$ and $^{239}_{94}Pu$

***Knowledge Required:*** The definition of "isotope," and the associated symbolic representation.

***Thinking it Through:*** Isotopes are forms of the same element that differ only in the number of neutrons in their nuclei. The identity of an element is determined by the number of protons (the atomic number), which is represented by the left-hand subscript. Choice (**B**), the only choice with the same element symbol for both atoms, also has the identical atomic number (92). The other three choices have identical mass numbers, which simply means that the sum of protons plus neutrons is the same for each one. *Note*: The similar terms, *isotope, isomer,* and *allotrope* are often confused.

---

**AS-4.**    Which particle, if lost from the nucleus, will **not** result in a change in the atomic number?

(A)    proton

(B)    alpha particle

(C)    beta particle

(D)    neutron

***Knowledge Required:*** (1) The general composition of atomic nuclei (numbers of neutrons and protons). (2) The modes of radioactive decay available to nuclei. (3) The meanings of the terms alpha particle and beta particle.

***Thinking it Through:*** Any of these particles could be lost from the nucleus during a nuclear (not chemical) change. Since the atomic number is determined by the number of positively charged protons in the nucleus, emission of *any* charged particle would change the atomic number. A proton, choice (**A**), has one unit of positive charge. An alpha particle, choice (**B**), has two units of positive charge. A beta particle, choice (**C**), has one unit of negative charge. The only remaining choice is the neutron, which has no charge, and therefore does not change the atomic number.

**AS-5.** When alpha particles were shot at a metal foil target to probe the structure of the atom, most of the particles passed through without any deflection in path. Some particles were deflected at large angles. This indicated to Rutherford that

(A) the metal foil was continuous matter.

(B) the mass of the metal atoms was spread out evenly.

(C) the atoms of the metal were mostly empty space.

(D) the alpha particles had great penetrating power.

*Knowledge Required:* The general features of Ernest Rutherford's alpha-particle-scattering experiment.

*Thinking it Through:* Prior to Rutherford's classic scattering experiment, atoms were thought to consist of a cloud of positive charge with electrons embedded throughout. Rutherford discovered that some alpha particles were scattered at large angles. The only explanation was that most of the mass and all of the positive charge had to be concentrated in a tiny fraction of the volume of atom. The rest of the atom had to be mostly empty space. Choice (B) describes atomic theory *before* Rutherford's experiment. Choice (A) suggests that atoms do not exist at all. Choice (D) does not bear directly on Rutherford's experiment. The large scattering angles of the alpha particles were *only* consistent with an atomic model in which the positive charge and almost all of the mass were concentrated in a small, dense center. The rest of the atom had to be mostly empty space, choice (C).

**AS-6.** An atom has a valence shell electron configuration of $ns^1$. To which group of elements in the periodic table does it belong?

(A) transition elements

(B) lanthanides

(C) alkaline earth metals

(D) alkali metals

*Knowledge Required:* (1) The symbolic representation of electron configuration. (2) The meaning of the term "valence shell." (3) The names associated with particular families or periods in the periodic table. (4) The ability to correlate symbolic representations with locations in the periodic table.

*Thinking it Through:* Elements having $ns^1$ as the valence-shell-electron configuration have a single electron in the $s$ orbital. Such elements include H ($1s^1$), Li ($2s^1$), Na ($3s^1$), and K ($4s^1$). These are the ***alkali metals***, found in Group IA of the periodic table, choice (D). The alkaline earth metals, Group IIA, all have electron configurations of $ns^2$, choice (C). Transition elements, choice (A), and lanthanides, choice (B), have varying numbers of valence electrons. Their characteristic valence electron configurations are $ns^2(n-1)d^x$ and $ns^2(n-1)d^x(n-2)f^y$ respectively. The maximum value of $x$ is 10 and the maximum value of $y$ is 14.

**AS-7.** The energy of a photon is greatest in the case of

(A) X-rays.

(B) ultraviolet radiation

(C) visible light.

(D) infrared radiation.

***Knowledge Required:*** (1) The meaning of the term "photon." (2) The names associated with regions of the electromagnetic spectrum. (3) The qualitative knowledge of the relationships between energy, frequency, and wavelength ( $E = h\nu = \dfrac{hc}{\lambda}$ ).

***Thinking it Through:*** A photon is a packet of energy, the quantity of which is related to the frequency of the electromagnetic radiation. *Higher* frequencies ($\nu$) and *shorter* wavelengths ($\lambda$) correlate with higher photon energies. All electromagnetic radiation travels at the velocity of light ($c$). The labels associated with regions of the electromagnetic spectrum are shown in the schematic diagram. Visible light and infrared radiation, choices (**C**) and (**D**), are relatively low energy. Ultraviolet radiation, choice (**B**), is higher in energy, but X-rays, choice (**A**), are highest.

**Electromagnetic Spectrum**

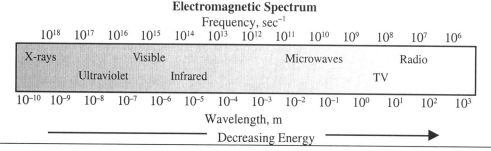

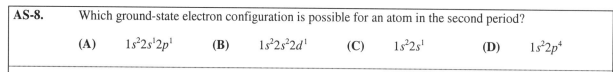

---

**AS-8.** Which ground-state electron configuration is possible for an atom in the second period?

    (**A**)     $1s^2 2s^1 2p^1$     (**B**)     $1s^2 2s^2 2d^1$     (**C**)     $1s^2 2s^1$     (**D**)     $1s^2 2p^4$

---

***Knowledge Required:*** (1) The symbolic representation of electron configuration. (2) The meaning of the term "second period." (3) The allowed values for principal and angular momentum quantum numbers. (4) The order of filling of electron orbitals. (5) The ability to correlate symbolic representations with positions in the periodic table.

***Thinking it Through:*** For the second period, the principal quantum number, $n$, is equal to 2. For a principal quantum number of 2, the angular momentum (shape) quantum number, $l$, has allowed values of 0 or 1. The letter $s$ is associated with $l = 0$, and the letter $p$ is associated with $l = 1$. Choice (**B**) cannot be correct, since only $s$ and $p$ orbitals are possible when $n = 2$. Choice (**D**) is eliminated because the $2s$ orbital must fill before the $2p$ orbital. The $2s$ orbital has room for two electrons, which must be occupied before the $2p$ orbital begins to fill. Choice (**A**) is eliminated because the order of filling is incorrect. Choice (**C**) has the correct orbitals present for a second period element, and also displays the correct order for filling orbitals. It is the electron configuration for lithium.

---

**AS-9.** The number of unpaired electrons in a gaseous selenium atom is

    (**A**)     2     (**B**)     3     (**C**)     4     (**D**)     5

*Knowledge Required:* (1) Hund's rule. (2) The order of filling of electron orbitals. (3)The ability to correlate symbolic representations with positions in the periodic table.

*Thinking it Through:* Selenium is in Group VIA, the fourth column of the *p*-block. Lower energy orbitals are completely filled. Hund's rule specifies that electrons in orbitals of the same energy do not pair until they have to. Initially, there are three equal-energy *p* orbitals. The order of filling first places a single electron in each *p* orbital, and then forming one pair with the fourth electron. This leaves *two* electrons unpaired in equal-energy orbitals. Here is a symbolic representation of the same information.

Se $[Ar]^{18} 3d^{10} 4s^2 4p^4$

$$3d^{10} \qquad 4s^2 \qquad 4p^4$$

Note: *Paramagnetism* is an experimental phenomena associated with atoms having unpaired electrons. Such substances are attracted to a magnetic field.

---

**AS-10.** Which electron transition in a hydrogen atom is associated with the largest *emission* of energy?

(A)  $n = 2$ to $n = 1$  (B)  $n = 2$ to $n = 3$

(C)  $n = 2$ to $n = 4$  (D)  $n = 3$ to $n = 2$

*Knowledge Required:* (1) The nature of quantized energy transfer between energy states. (2) The relative spacing of energy levels from $n = 1$ to $n = \infty$. (3) The interpretation of energy level changes.

*Thinking it Through:* Atomic line spectra provide evidence that the energy state of an electron in an atom is quantized. Only transitions between discrete energy states can take place.

Energy must be *absorbed* by the atom for an electron to move from one energy state in an atom to another energy state that is more remote from the nucleus. This observation eliminates both choices (**B**) and (**C**). The transitions in (**A**) and (**D**) both emit energy, and involve movement to the next lower energy state. Examining the relative spacing of energy levels reveals that $n = 2$ to $n = 1$ is the greater of the two.

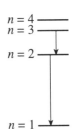

# Practice Questions

1. In all neutral atoms, there are equal numbers of

   (A)  protons and neutrons.  (B)  positrons and electrons.

   (C)  neutrons and electrons.  (D)  electrons and protons.

2. Which element is represented by $^{56}_{24}X$ ?

   (A)  iron  (B)  germanium  (C)  barium  (D)  chromium

3. Which statement concerning the structure of the atom is correct?

   (A)  Protons and neutrons have most of the mass and occupy most of the volume of the atom.

   (B)  Electrons have most of the mass and occupy most of the volume of the atom.

   (C)  Electrons have most of the mass but occupy very little of the volume of the atom.

   (D)  Protons and neutrons have most of the mass but occupy very little of the volume of the atom.

4. The number of neutrons in the $^{37}_{17}Cl^-$ ion is

    (A)      17      (B)      20      (C)      21      (D)      37

5. An argon atom is isoelectronic with

    (A)      Cl      (B)      Ca      (C)      $Ti^{4+}$      (D)      $N^{3-}$

6. Which pair of particles has the same number of electrons?

    (A)      $F^-$, $Mg^{2+}$                    (B)      Ne, Ar

    (C)      $Br^-$, Se                  (D)      $Al^{3+}$, $P^{3-}$

7. Which ion has twenty-six electrons?

    (A)      $Cr^{2+}$      (B)      $Fe^{2+}$      (C)      $Ni^{2+}$      (D)      $Cu^{2+}$

8. Which statement is true?

    (A)      The nucleus of an atom contains neutrons and electrons.

    (B)      The atomic number of an element is the number of protons in one atom.

    (C)      The mass number of an atom is the number of protons in the nucleus plus the number of electrons outside.

    (D)      The number of electrons outside the nucleus is the same as the number of neutrons in the nucleus.

9. In what respect does an atom of magnesium, Mg, differ from a magnesium ion, $Mg^{2+}$?

    (A)      The ion has an inert gas electron configuration; the atom does not.

    (B)      The positive charge on the nucleus of the ion is two units greater than the nuclear charge on the atom.

    (C)      The ion has two more protons than the atom.

    (D)      The ion has two more planetary electrons than the atom.

10. A sodium ion differs from a sodium atom in that the sodium ion

    (A)      has fewer electrons.

    (B)      is an isotope of sodium.

    (C)      exists only in solution.

    (D)      has a negative charge on its nucleus.

11. Which term best characterizes the relation of hydrogen to deuterium?

    (A)      allotropes      (B)      isomers      (C)      isotopes      (D)      polymers

12. The element **X** occurs naturally to the extent of 20.0% $^{12}$**X** and 80.0% $^{13}$**X**. The atomic mass of **X** is nearest

    (A)      12.2      (B)      12.5      (C)      12.8      (D)      13.0

13. In which pair are the two species *both* isoelectronic and isotopic?

(A) $^{40}_{20}Ca^{2+}$ and $^{40}_{18}Ar$    (B) $^{39}_{19}K^+$ and $^{40}_{19}K^+$

(C) $^{24}_{12}Mg^{2+}$ and $^{25}_{12}Mg$    (D) $^{56}_{26}Fe^{2+}$ and $^{57}_{26}Fe^{3+}$

14. An atom of the element of atomic number 84 and mass number 199 emits an alpha particle. The residual atom after this change has an atomic number of ____ and a mass number of _____.

(A) 82, 195    (B) 84, 203    (C) 85, 195    (D) 86, 199

15. In effecting nuclear changes by bombarding target nuclei with positively charged alpha particles, it is necessary to accelerate these particles to high speed because it is necessary to

(A) drive the positive particles through the electron cloud.

(B) overcome the force of repulsion of the nucleus.

(C) focus the bombarding particle more accurately.

(D) strip electrons from the atom.

16. What significant information about atomic structure came from the Millikan experiment using charged oil drops?

(A) Millikan showed that cathode rays were identical to a stream of electrons coming from an atom.

(B) Millikan confirmed that the neutron and proton were of about the same mass.

(C) Millikan determined the magnitude of the charge on an electron.

(D) Millikan proved that the mass of an atom was concentrated in the nucleus.

17. What is the valence electron configuration for the element in Period 5, Group 3A?

(A) $5s^25p^1$    (B) $3s^23p^5$    (C) $3s^23p^3$    (D) $5s^25p^3$

18. Which electron configuration is *impossible*?

(A) $1s^22s^22p^63s^2$    (B) $1s^22s^22p^62d^2$

(C) $1s^22s^22p^63s^23p^6$    (D) $1s^22s^22p^53s^1$

19. The ground–state electronic configuration of the manganese atom, Mn, is

(A) $1s^22s^22p^63s^23p^64s^24d^5$    (B) $1s^22s^22p^63s^23p^63d^7$

(C) $1s^22s^22p^63s^23p^64s^24p^5$    (D) $1s^22s^22p^63s^23p^63d^54s^2$

20. Which species has this ground-state electron arrangement?

$$1s^22s^22p^63s^23p^63d^{10}$$

(A) Ni    (B) $Ni^{2+}$    (C) Zn    (D) $Zn^{2+}$

21. The maximum number of electrons that can occupy an orbital labeled $d_{xy}$ is

(A) 1    (B) 2    (C) 3    (D) 4

22. An atom of Fe has two $4s$ electrons and six $3d$ electrons. How many *unpaired* electrons would there be in the $Fe^{2+}$ ion?

    (A)    one      (B)    two      (C)    three      (D)    four

23. Which of these species (is/are) paramagnetic?

                     $Ti^{4+}$             $Fe^{2+}$          $Zn^{o}$

    (A)    $Fe^{2+}$ only                  (B)    $Zn^{o}$ only

    (C)    $Ti^{4+}$ and $Fe^{2+}$ only       (D)    $Fe^{2+}$ and $Zn^{o}$ only

24. Which of these electron diagrams could represent the ground state of the $p$ valence electrons of carbon?

    (A)                      (B)   ↑   ↓   —

    (C)                      (D)   ↑   ↑   —

25. The existence of discrete (quantized) energy levels in an atom may be inferred from

    (A)    experiments on the photoelectric effect.

    (B)    diffraction of electrons by crystals.

    (C)    X-ray diffraction by crystals.

    (D)    atomic line spectra.

26. Which emission line in the hydrogen spectrum occurs at the highest frequency?

    (A)    $n = 3 \rightarrow n = 1$           (B)    $n = 4 \rightarrow n = 2$

    (C)    $n = 7 \rightarrow n = 5$           (D)    $n = 10 \rightarrow n = 8$

27. Which set of quantum numbers is correct and consistent with $n = 4$?

    (A)   $l = 3$     $m_l = -3$     $m_s = +\frac{1}{2}$     (B)   $l = 4$     $m_l = +2$     $m_s = -\frac{1}{2}$

    (C)   $l = 2$     $m_l = +3$     $m_s = +\frac{1}{2}$     (D)   $l = 3$     $m_l = -3$     $m_s = +1$

28. When an atom of an electropositive atom becomes an ion it

    (A)    gains electrons.              (B)    becomes larger.

    (C)    emits an alpha particle.       (D)    does none of these.

29. The orbitals of $2p$ electrons are often represented as being

    (A)    elliptical.                   (B)    tetrahedral.

    (C)    dumbbell shaped.         (D)    spherical.

30. Helium, $^4_2He$, has two electrons in the $1s$ orbital. When it becomes singly ionized, forming $He^+$,

    (A)    its spectrum resembles that of the hydrogen spectrum.

    (B)    the remaining electron is easier to remove.

    (C)    the nuclear charge has decreased by one.

    (D)    it has lost one atomic mass unit.

## *Answers to Study Questions*

1. A
2. C
3. B
4. D
5. C
6. D
7. A
8. C
9. A
10. A

## *Answers to Practice Questions*

1. D
2. D
3. D
4. B
5. C
6. A
7. C
8. B
9. A
10. A
11. C
12. C
13. B
14. A
15. B
16. C
17. A
18. B
19. D
20. D
21. B
22. D
23. A
24. D
25. D
26. A
27. A
28. D
29. C
30. A

# Molecular Structure and Bonding

Understanding molecular structure requires knowledge of the spatial arrangement of atoms in a molecule. The geometric arrangement of atoms in molecules can be determined by experiment, but bonding theories and properties such as dipole moment, which is a measure of molecular polarity, can help predict and rationalize molecular shape and chemical behavior. This section of most general chemistry courses deals with valence-shell electron-pair repulsion (VSEPR) models, prediction of bond angles, valence bond theory, hybrid orbitals, and molecular orbital theory.

## Study Questions

**MS-1.** Which molecule has exactly two unshared (lone) pairs of electrons on the central atom?

    **(A)**    $BF_3$      **(B)**    $OF_2$      **(C)**    $NF_3$      **(D)**    $XeF_2$

*Knowledge Required:* (1) The rules for drawing Lewis structures. (2) The exceptions to the octet rule.

*Thinking it Through:* Start by finding the total number of valence electrons that are available. The group number for the main group or representative (A group) elements gives this number. Fluorine, for example, is a member of Group VIIA and has 7 valence electrons. Distribute the electrons to satisfy the octet rule, keeping in mind the common exceptions.

$BF_3$    $3 + 3(7) = 24$ valence electrons    [Lewis structure]    No unshared pairs on boron.

$OF_2$    $6 + 2(7) = 20$ valence electrons    [Lewis structure]    Exactly two unshared pairs on oxygen.

$NF_3$    $5 + 3(7) = 26$ valence electrons    [Lewis structure]    One unshared pair on nitrogen.

$XeF_2$    $8 + 2(7) = 22$ valence electrons    [Lewis structure]    Three unshared pairs on xenon.

Note that there are two common exceptions here. Boron has too few valence electrons to form a full octet, and xenon can form an expanded octet, making both choices **(A)** and **(D)** incorrect. $NF_3$ has only one lone pair, so choice **(C)** is also incorrect. Only $OF_2$ has exactly two unshared pairs of electrons on the central atom, making choice **(B)** the correct answer.

**MS-2.** Consider the Lewis structure for $CH_3Cl$. What is the best description of the molecular shape?

    **(A)**    bent                  **(B)**    square

    **(C)**    square pyramidal      **(D)**    tetrahedral

*Knowledge Required:* (1) The interpretation of Lewis structures. (2) The valence-shell-electron-pair-repulsion (VSEPR) model for predicting molecular shape.

*Thinking it Through:* The first step in applying the VSEPR method is to count the number of electron regions around the central atom. In this case, the Lewis structure is given, showing four pairs of electrons around the carbon atom. These electron pairs are all used to form bonds, and there are no lone pairs. Electron pair repulsion is minimized when the four electron pairs form a tetrahedron around the carbon atom, choice (**D**). Observe the similarity to methane, $CH_4$, which was probably the first tetrahedral molecule you studied. *Note:* In other questions you may be asked about the *tetrahedral bond angle*, which is 109.5°.

---

**MS-3..** The fact that $BCl_3$ is a planar molecule while $NCl_3$ is pyramidal can be explained several different ways. Which statement is the best rationalization?

   (**A**)   Nitrogen is more electronegative than boron.

   (**B**)   The nitrogen atom in $NCl_3$ has a lone pair of electrons, whereas the boron atom in $BCl_3$ does not.

   (**C**)   The nitrogen atom is smaller than the boron atom.

   (**D**)   The boron atom in $BCl_3$ is $sp^3$ hybridized, while the nitrogen atom in $NCl_3$ is $sp^2$ hybridized.

...............................................................................................................................................

*Knowledge Required:* (1) The valence-shell-electron-pair-repulsion (VSEPR) model for predicting molecular shape. (2) The valence-bond theory for predicting hybridization of orbitals.

*Thinking it Through:* Choices (**A**) and (**C**) are both true statements, but neither provides the *reason* for the difference in geometries. The VSEPR model does provide a rationalization for the difference. Choice (**D**) has the hybridization reversed. $NCl_3$ has an $sp^3$ hybridized nitrogen atom, and $BCl_3$ has an $sp^2$ hybridized boron atom. Choice (**B**), the correct answer, recognizes that boron has only enough valence electrons to form three bonds, with no lone pairs. The six electrons arrange themselves to have 120° Cl–B–Cl bond angles. On the other hand, when $NCl_3$ forms three bonds it has an electron pair left over to complete the octet. The four electron pairs arrange themselves in the shape of a tetrahedron around the nitrogen atom. Here are the Lewis structures and their three-dimensional representations:

---

**MS-4.** According to the VSEPR model, the geometric structure of $H_2O$ is

   (**A**)   bent at an angle of 104.5° because of the greater repulsion of two lone pairs relative to that of the two bonding pairs.

   (**B**)   bent at an angle of 120° because of the mutual repulsion of the six valence electrons on oxygen.

   (**C**)   bent at an angle of 90° because of the perpendicular relationship of the oxygen $p$ orbitals relative to each other.

   (**D**)   linear with a bond angle of 180° because of the mutual repulsion of the bonding pairs of electrons.

*Knowledge Required:* (1) Valence-shell-electron-pair-repulsion (VSEPR) model for predicting molecular shape. (2) Effect of lone pairs on predicted bond angles.

*Thinking it Through:* Recall or write the Lewis structure for the water molecule.

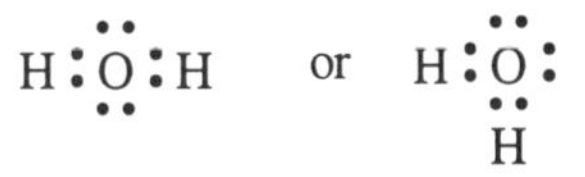

$$H:\overset{..}{\underset{..}{O}}:H \quad \text{or} \quad H:\overset{..}{\underset{\underset{H}{|}}{O}}:$$

A Lewis structure by itself does not provide information about geometry. Use VSEPR to predict geometry by counting bonded and nonbonded electron pairs. If there are four bonded pairs to identical atoms, the exact tetrahedral angle of 109.5° can be expected. That is the case with $CH_4$. In $NH_3$ there are three bonded pairs and one nonbonded pair, so the bond angle closes slightly (to 107°) as the result of the lone pair requiring more space than the bonded pairs. In the water molecule, with two nonbonded pairs, the H–O–H bond angle closes to 104.5° because of repulsion from the lone pairs. (**A**) is the correct choice. Choice (**B**) has the bond angle larger, rather than smaller, than the tetrahedral angle. Choice (**C**) treats the orbitals as if they had the geometry of atomic orbitals. Choice (**D**) fails to recognize the three-dimensional nature of molecules.

---

**MS-5.** Sulfur dioxide can be described by the structures shown. This symbolism indicates that the

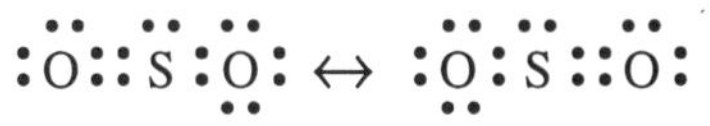

$$:\overset{..}{O}::\overset{..}{S}:\overset{..}{\underset{..}{O}}: \leftrightarrow :\overset{..}{\underset{..}{O}}:\overset{..}{S}::\overset{..}{O}:$$

**(A)** two bonds in $SO_2$ are of equal length, and the electronic distribution in the two sulfur-to-oxygen bonds is identical.

**(B)** single bond is longer than the double bond, and the electronic distribution in the two sulfur-to-oxygen bonds differs.

**(C)** electron pair in the $SO_2$ molecule alternates back and forth between the two sulfur-oxygen electron pairs, so that the two different bonds seem to exchange positions.

**(D)** $SO_2$ molecule revolves, so that the two different sulfur-to-oxygen bonds seem to exchange positions.

*Knowledge Required:* (1) The meaning of resonance structures. (2) The interpretation of Lewis structures.

*Thinking it Through:* A single Lewis electron-dot formula cannot always adequately describe bonding between atoms. If two (or more) formulas satisfy the octet rule, the Lewis structures are both written and resonance is indicated with a double-headed arrow. This does *not* mean that the molecule oscillates between the structures. Rather, it means that the actual structure is intermediate between (or among) the resonance forms. Choices (**C**) and (**D**) suggest movement of electrons or atoms, which does not occur. Choice (**B**) suggests that the sulfur–oxygen bonds are not equivalent, but they are. Correct choice (**A**) recognizes that the two bonds are identical. The need to draw two forms reflects a limitation of the Lewis-dot model.

---

**MS-6.** A compound consisting of an element having a low ionization energy and a second element having a high electron affinity is likely to have

**(A)** covalent bonds. **(B)** metallic bonds.

**(C)** coordinate covalent bonds. **(D)** ionic bonds.

*Knowledge Required:* (1) The meaning of the terms ionization energy and electron affinity. (2) The characteristics of different types of chemical bonds. (3) The relationship between types of chemical bonds and values of ionization energy and electron affinity.

*Thinking it Through:* Elements with low ionization energy readily lose electrons to form positive ions. Elements with high electron affinity readily accept electrons to form negative ions. Ionic bonds, correct choice (**D**), result when atoms exchange electrons. Covalent bonds, choice (**A**), result when atoms each contribute an electron to a shared pair. Coordinate covalent bonds, choice (**C**), are similar except one of the two atoms furnishes both electrons. Metallic bonds, choice (**B**), result when atoms free one or more valence electrons to the metal lattice.

---

**MS-7.** The compound $CF_3CHClF$ is being considered as a replacement for $CBrF_3$ as a fire-extinguishing agent because $CBrF_3$ has been shown to deplete stratospheric ozone. What are the most probable products if a molecule of $CF_3CHClF$ is bombarded with high-energy photons?

| Bond | Bond Energy, $kJ \cdot mol^{-1}$ |
|------|------|
| C–C | 346 |
| C–Cl | 327 |
| C–F | 485 |
| C–H | 411 |

(**A**)     $CF_3 + CHClF$

(**B**)     $CF_2CHClF + F$

(**C**)     $CF_3CHF + Cl$

(**D**)     $CF_3CClF + H$

.................................................................................................................

*Knowledge Required:* (1) The knowledge of the interaction of photons with covalent bonds. (2) The interpretation of bond energy data.

*Thinking it Through:* High-energy photons, such as those in the ultraviolet part of the electromagnetic spectrum, have enough energy to disrupt covalent bonds. In choosing from the set of possible products resulting from breaking at least one covalent bond, it may be helpful to consider the Lewis structure. Choice (**A**) requires breaking the C–C bond, which would require 346 kJ per mol. Choice (**B**) requires breaking one of the C–F bonds, but that would require 485 kJ per mol. The correct choice (**C**) results from breaking a C–Cl bond, the bond requiring the least energy of the choices given, 327 kJ per mol. Choice (**D**) could result if a C–H bond were broken, but that is less favorable at 411 kJ per mol. Note: Halon compounds contain bromine covalently bonded to carbon. This bond is even more easily broken by high-energy photons. The C–Br bond energy is $285 \; kJ \cdot mol^{-1}$.

---

**MS-8.** A simple method of showing experimentally that a solid substance may be ionic is to show that it

(**A**)     has a high melting point.

(**B**)     is soluble in polar solvents.

(**C**)     depresses the freezing point of water.

(**D**)     conducts current when dissolved in water.

.................................................................................................................

*Knowledge Required:* (1) The nature of ionic bonding and properties of ionic compounds. (2) The attributes of solutions of ionic compounds in water.

*Thinking it Through:* Choice (**A**) could be true, but melting point is not a valid criterion for establishing the presence of ionic bonds. Several covalent network solids have melting points of several thousand degrees Celsius. Choice (**B**) is true for ionic compounds, as it is for polar covalent compounds that do not react with water to form ions. Alcohols, for example, are miscible with water, but they are not ionic compounds. Choice (**C**) is true for both ionic and covalent compounds. In fact, *any* nonvolatile solute that is dissolved in water depresses its freezing point. Ethylene glycol, a covalent compound, is used as antifreeze in automobile radiators. Choice (**D**) is true only for ionic compounds. (Some liquid and gaseous covalent compounds react with water to form ions; hence, the necessity of including the word "solid" in the stem.)

**MS-9.** Which of these is a nonpolar molecule?

(A)
$$\begin{array}{c} H \\ \diagdown \\ C=O \\ \diagup \\ H \end{array}$$

(B)
$$\begin{array}{c} H-O \\ \diagdown \\ H \end{array}$$

(C)
$$\begin{array}{c} F \qquad F \\ \diagdown \quad \diagup \\ C=C \\ \diagup \quad \diagdown \\ F \qquad F \end{array}$$

(D) I–Br

.................................................................................................................................................................

*Knowledge Required:* (1) The relative electronegativity of atoms. (2) The properties of polar bonds. (3) The molecular geometry of molecules. (4) The relationship of dipole moment to molecular geometry.

*Thinking it Through:* For a molecule to be nonpolar when it has polar bonds, it must be symmetric enough that all the bond dipoles cancel each other out. For choice (A), the pull of the oxygen on the electrons shared with carbon is not cancelled from the other side, and the molecule is polar. The same is true for choices (B) and (D). Choice (C) contains polar carbon-fluorine bonds, with the polarity of each one cancelled by a carbon–fluorine bond on the opposite side of the molecule.

---

**MS-10.** For which molecule can the bonding be described in terms of $sp^3$ hybrid orbitals of the central atom?

(A)   $SF_6$     (B)   $BF_3$     (C)   $PCl_5$     (D)   $NH_3$

.................................................................................................................................................................

*Knowledge Required:* (1) Geometry of hybrid orbitals. (2) Number of valence electrons for several atoms.

*Thinking it Through:* To form $sp^3$ hybrid orbitals, valence-bond theory uses one $s$ orbital and three $p$ orbitals to produce four equivalent orbitals oriented toward the corners of a tetrahedron. Each of the hybrid orbitals has a capacity of two electrons, for a maximum of eight electrons in the valence shell. When sulfur forms two-electron bonds with each of six fluorine atoms, as in choice (A), the sulfur must form a hybrid that uses *six* atomic orbitals rather than four. Similarly, the phosphorus atom in choice (C) must form hybrids that use *five* atomic orbitals. Boron, with only three valence electrons, cannot place one in each of four hybrid orbitals. It is only able to form $sp^2$ hybrid orbitals, except when a bonding partner provides both of the share electrons. Boron is a $sp^2$ hybrid in $BF_3$, choice (B). A nitrogen atom, with five valence electrons, forms bonds with three hydrogen atoms. There are eight valence electrons, enough to fill four $sp^3$ hybrid orbitals, choice (D).

## Practice Questions

1. $NH_3$ (pyramidal geometry) reacts with $BF_3$ (planar geometry) to form the addition compound, $H_3NBF_3$. What is the geometry around the nitrogen and boron centers in the addition compound?

   (A) Both centers are tetrahedral.     (B) Nitrogen is tetrahedral; boron is linear.

   (C) Nitrogen is pyramidal; boron is planar.     (D) Nitrogen is planar; boron is pyramidal.

2. The molecule of the type $ML_4$ consists of four single bonds and no lone pairs. What structure is it expected to assume?

    (A) square planar     (B) trigonal planar

   (C) trigonal pyramidal     (D) tetrahedral

3. The shape that most closely describes the $NF_3$ molecule is

   (A)   octahedral.                                (B)   trigonal planar.

   (C)   trigonal pyramidal.                         (D)   tetrahedral.

4. In which pair are the molecules geometrically similar?

   (A)   $SO_2$ and $CO_2$                            (B)   $PH_3$ and $BF_3$

   (C)   $CO_2$ and $OF_2$                            (D)   $SO_2$ and $O_3$

5. Which is planar?

   (A)   $NH_3$        (B)   $SO_3^{2-}$        (C)   $CO_3^{2-}$        (D)   $CCl_4$

6. Which is linear?

   (A)   $H_2S$        (B)   $NH_3$        (C)   $NO_2$        (D)   $CO_2$

7. Consider the given Lewis structure for $BrF_5$. What is the predicted shape for the molecule as a whole?

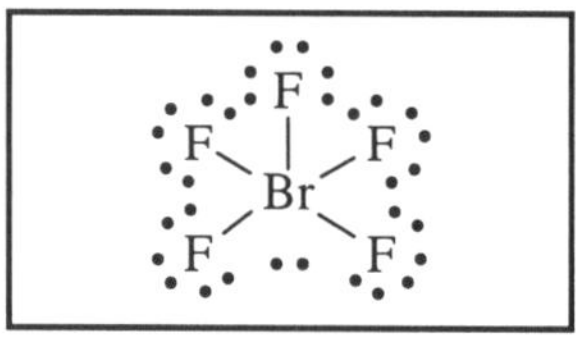

   (A)   square pyramidal                            (B)   trigonal bipyramidal

   (C)   trigonal pyramidal                          (D)   octahedral

8. What is the shape of the $XeF_4$ molecule?

   (A)   square planar                               (B)   trigonal bipyramidal

   (C)   tetrahedral                                 (D)   trigonal pyramidal

9. What set of species is arranged in order of increasing O–N–O bond angle?

   (A)   $NO_2^-$, $NO_2$, $NO_2^+$                  (B)   $NO_2$, $NO_2^-$, $NO_2^+$

   (C)   $NO_2^+$, $NO_2$, $NO_2^-$                  (D)   $NO_2$, $NO_2^+$, $NO_2^-$

10. Which has the largest bond angle?

   (A)   angle O–S–O in $SO_4^{2-}$                  (B)   angle Cl–C–Cl in $HCCl_3$

   (C)   angle F–Be–F in $BeF_2$                     (D)   angle H–O–H in $H_2O$

11. The structure of the $CO_3^{2-}$ ion can be described in the Lewis formulation by these structures. This means that

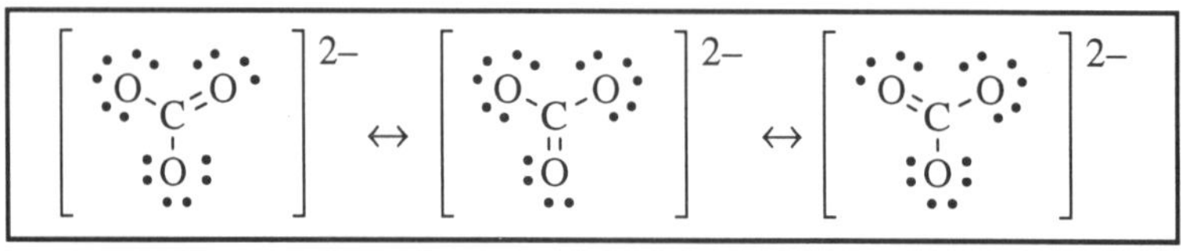

   (A)   two carbon-to-oxygen bonds are single bonds; the third is a double bond.

   (B)   three independent forms of the $CO_3^{2-}$ ion coexist in equilibrium.

   (C)   the electrons must be rapidly exchanging among the three forms.

   (D)   the $CO_3^{2-}$ ion exists in only one form: an average of the three principal structures shown.

**12.** Which concept describes the formation of four equivalent, single, covalent bonds by carbon in its compounds that resemble methane, $CH_4$?

(A)  hydrogen bonding

(B)  hybridization

(C)  sigma bonding

(D)  coordinate covalent bonding

**13.** Which type of hybrid orbital is used in $CO_2$?

(A)  $sp$

(B)  $sp^2$

(C)  $sp^3$

(D)  $dsp^3$

**14.** Which compound would be expected to have the largest dipole moment?

(A)  $CO_2$ (linear)

(B)  $SO_2$ (bent)

(C)  $BF_3$ (trigonal planar)

(D)  $CF_4$ (tetrahedral)

**15.** The O–Si–O bond angles in $SiO_2$ (quartz) are closest to

(A)  180°

(B)  120°

(C)  110°

(D)  100°

**16.** The molecule $\ddot{\text{O}}=\text{C}=\ddot{\text{N}}-\text{H}$ has been detected in gas clouds between stars. The predicted C–N–H bond angle is about

(A)  90°

(B)  109°

(C)  120°

(D)  180°

**17.** Knowing that F is more electronegative than either B or P, what conclusion can be drawn from the fact that $BF_3$ has no dipole moment, but $PF_3$ does?

(A)  $BF_3$ is not spherically symmetrical, but $PF_3$ is.

(B)  The $BF_3$ molecule must be trigonal planar.

(C)  The $BF_3$ molecule must be linear.

(D)  The atomic radius of P is larger than the atomic radius of B.

**18.** How many valence electrons are represented in the Lewis electron-dot structure for $SO_2$?

(A)  6

(B)  8

(C)  18

(D)  32

**19.** Molecules of which compounds violate the octet rule?

| $NO_2$ | $CH_2Cl_2$ | $XeF_4$ | $NCl_3$ |
|--------|------------|---------|---------|
| 1      | 2          | 3       | 4       |

(A)  1 and 2

(B)  1 and 3

(C)  2 and 3

(D)  2 and 4

**20.** The boiling point of $H_2O$, compared with the other members of the series, can be explained by

| **Boiling Points of Group VIA Hydrides** | | | |
|---------|---------|----------|---------|
| $H_2O$  | $H_2S$  | $H_2Se$  | $H_2Te$ |
| 100 °C  | –61 °C  | –45 °C   | –2 °C   |

(A)  London dispersion forces.

(B)  dipole—induced dipole forces.

(C)  hydrogen bonding.

(D)  nonpolar covalent bonding.

**21.** The fact that $Pt(NH_3)_2Cl_2$ exists in two different isomeric forms offers evidence that the geometry is

(A)    octahedral.  (B)    square planar.

(C)    tetrahedral.  (D)    trigonal planar.

**22.** Which species has *both* covalent and ionic bonds?

(A)    $NH_3BF_3$  (B)    $H_3O^+$  (C)    NaKS  (D)    $Mg(CN)_2$

**23.** When the carbon–carbon bonds in ethane $(C_2H_6)$, ethene $(C_2H_4)$, and benzene $(C_6H_6)$, are arranged in order of increasing length (shortest bonds first), what is the correct order?

(A)    $C_2H_6 < C_2H_4 < C_6H_6$  (B)    $C_2H_4 < C_2H_6 < C_6H_6$

(C)    $C_6H_6 < C_2H_4 < C_2H_6$  (D)    $C_2H_4 < C_6H_6 < C_2H_6$

**24.** These two electron-dot formulas for carbon dioxide both satisfy the octet rule, but one is preferred over the other. Which of the statements identifies the preferred structure and also the reason it is preferred?

Structure **1**          Structure **2**

(A)    Structure **1** is preferred because the triple bond is stronger than the double bond.

(B)    Structure **1** is preferred because the formal charge on carbon is negative.

(C)    Structure **2** is preferred because the formal charge on each atom is zero.

(D)    Structure **2** is preferred because the bonds are equal.

**25.** Which element is most likely to form a triple bond?

(A)    Pb  (B)    F  (C)    N  (D)    S

**26.** Which is an isomer of

?

(A)

(B)

(C)

(D)

27. How many sigma bonds and how many pi bonds are represented in this structure?

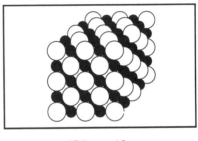

|  | **Sigma Bonds** | **Pi Bonds** |
|---|---|---|
| (A) | 6 | 3 |
| (B) | 7 | 2 |
| (C) | 8 | 1 |
| (D) | 9 | 0 |

28. Sodium chloride, NaCl, crystallizes in a face-centered cubic lattice of chloride ions, with the smaller sodium ions occupying holes between the chloride ions. How many $Cl^-$ ions are in contact with any single $Na^+$ ion?

(A)   4        (B)   6        (C)   8        (D)   12

29. Resonance structures describe molecules that have

(A)   hybrid orbitals.

(B)   rapid equilibria.

(C)   resonating electrons.

(D)   multiple electron-dot formulas.

30. The complete Lewis structure of $\overset{:O:}{\underset{:Cl \diagdown C \diagup Cl:}{|}}$ will have

(A)   at least one lone pair on each atom.

(B)   at least one double bond.

(C)   both polar and nonpolar bonds.

(D)   resonance forms.

## Answers to Study Questions

| | | |
|---|---|---|
| 1. B | 5. A | 9. C |
| 2. D | 6. D | 10. D |
| 3. B | 7. C | |
| 4. A | 8. D | |

## Answers to Practice Questions

| | | |
|---|---|---|
| 1. A | 11. D | 21. B |
| 2. D | 12. B | 22. D |
| 3. C | 13. A | 23. D |
| 4. D | 14. B | 24. C |
| 5. C | 15. C | 25. C |
| 6. D | 16. C | 26. A |
| 7. A | 17. B | 27. A |
| 8. A | 18. C | 28. B |
| 9. A | 19. B | 29. D |
| 10. C | 20. C | 30. B |

# *Stoichiometry*

Chemical stoichiometry is the area of study that considers the quantities of materials in chemical formulas and equations. Quite simply, it is chemical arithmetic. The word itself is derived from *stoicheion*, the Greek word for element and *metron*, the Greek word for measure. When based on chemical formulas, stoichiometry is used to convert between mass and moles, to calculate the number of atoms, to calculate percent composition, and to interpret the mole ratios expressed in a chemical formula. Most topics in chemical arithmetic depend on the interpretation of balanced chemical equations. Mass/mole conversions, calculation of limiting reagent and percent yield, and various relationships among reactants and products are commonly included in this topic area.

## *Study Questions*

| ST-1. | What is the empirical formula for the substance with this analysis? | Elemental Analysis by Mass | |
|---|---|---|---|
| | | Na | 54.0% |
| | | B | 8.50% |
| | | O | 37.5% |

| (A) | $Na_3BO_3$ | (B) | $Na_2B_2O_3$ | (C) | $Na_4BO_4$ | (D) | $Na_3B_2O_2$ |
|---|---|---|---|---|---|---|---|

*Knowledge Required:* (1) The meaning of empirical formula. (2) The interpretation of elemental analysis data.

*Thinking it Through:* An empirical formula is also called the simplest formula. Both terms refer to the lowest whole number ratios of atoms in a molecule or formula unit, and are derived from experiments. Such experiments yield the relative percentage of elements in a compound, expressed as percentage by mass. Formulas give the mole ratio of each element in the compound, so the first step is to convert each mass percentage to a relative mole value. It is easiest to do by assuming 100 g of the compound. Therefore, in each 100 grams of this compound there will be 54.0 g of sodium, 8.50 g of boron, and 37.5 g of oxygen. The molar mass of each element is used to find the number of moles.

$$54.0 \text{ g Na} \times \frac{1 \text{ mol Na}}{23.0 \text{ g Na}} = 2.35 \text{ mol Na} \qquad 8.50 \text{ g B} \times \frac{1 \text{ mol B}}{10.8 \text{ g B}} = 0.787 \text{ mol B} \qquad 37.5 \text{ g O} \times \frac{1 \text{ mol O}}{16.0 \text{ g O}} = 2.34 \text{ mol O}$$

These values correctly represent the ratio of sodium, boron, and oxygen atoms in the compound, but it is not the usual practice to write formulas such as $Na_{2.35}B_{0.787}O_{2.34}$. To find the ratio of whole numbers needed for the formula, divide each value by the smallest number of moles. In this case, the smallest number of moles is 0.787 mol B.

$$\frac{2.35 \text{ mol Na}}{0.787 \text{ mol B}} = \frac{2.99 \text{ mol Na}}{1.00 \text{ mol B}} \approx \frac{3 \text{ mol Na}}{1 \text{ mol B}} \qquad \frac{2.34 \text{ mol O}}{0.787 \text{ mol B}} = \frac{2.97 \text{ mol O}}{1.00 \text{ mol B}} \approx \frac{3 \text{ mol O}}{1 \text{ mol B}}$$

This whole number ratio shows that the correct formula is $Na_3BO_3$, which is choice (A).

*Note:* The percent composition values are experimental values, so do not expect the mole ratios to be *exactly* whole number ratios. However, if the mole ratios are more than one or two percent from whole numbers after the second step, consider multiplying through each ratio by a small integer to convert to whole numbers. See practice problem 2 for an example of this.

| ST-2. | What is the mass percent of oxygen in $Fe_2(SO_4)_3$? | Atomic Molar Mass Data | |
|---|---|---|---|
| | | Fe | 55.85 g·mol$^{-1}$ |
| | | S | 32.06 g·mol$^{-1}$ |
| | | O | 16.00 g·mol$^{-1}$ |

| (A) | 15.40% | (B) | 18.76% | (C) | 30.80% | (D) | 48.01% |
|---|---|---|---|---|---|---|---|

*Knowledge Required:* (1) The meaning of mass percent. (2) The interpretation of chemical formulas.

*Thinking it Through:* This type of question starts with a chemical formula and uses it to determine the mass percent for any element in a compound. Mass percent is a comparison of the mass of the element being considered to the molar mass of the compound. The ratio is then multiplied by 100 to make it a percent, which is the same as parts per hundred. In this case, the mass of oxygen present must be compared to the molar mass of the compound, $Fe_2(SO_4)_3$.

$$\% \text{ O} = \frac{\left(12 \text{ mol O} \times \frac{16.00 \text{ g}}{1 \text{ mol O}}\right)}{\left(2 \text{ mol Fe} \times \frac{55.85 \text{ g}}{1 \text{ mol Fe}}\right) + \left(3 \text{ mol S} \times \frac{32.06 \text{ g}}{1 \text{ mol S}}\right) + \left(12 \text{ mol O} \times \frac{16.00 \text{ g}}{\text{mol O}}\right)} \times 100$$

$$\% \text{ O} = \frac{192.0 \text{ g O}}{399.88 \text{ g Fe}_2(SO_4)_3} \times 100$$

$$\% \text{ O} = 48.01\% \text{ O by mass}$$

This is choice (**D**). Errors arise if the formula is misinterpreted and an incorrect number of atoms are counted, leading to an incorrect value for either the mass of oxygen or the mass of the total compound.

Mass percent questions are often part of a practical application, such as choosing an ore with the highest percentage of the metal desired or choosing a hydrocarbon fuel with the highest percentage of carbon. Note also that if a mass percent is given, it would be possible to work back to determine the mass of an element or a group of elements within a compound.

---

**ST-3.** A typical silicon chip, such as those in electronic calculators, has a mass of $2.3 \times 10^{-4}$ g. Assuming the chip is pure silicon, how many silicon atoms are in such a chip?

(**A**) $4.9 \times 10^{18}$ (**B**) $1.4 \times 10^{20}$ (**C**) $3.9 \times 10^{21}$ (**D**) $2.6 \times 10^{27}$

---

*Knowledge Required:* (1) The method for carrying out mass to mole conversions. (2) The application of Avogadro's number.

*Thinking it Through:* Avogadro's number, $6.02 \times 10^{23}$, gives the number of silicon atoms in each mole of silicon. This means that calculating the number of moles of silicon present is an essential first step before using Avogadro's number to find the number of silicon atoms.

$$2.3 \times 10^{-4} \text{ g} \times \frac{1 \text{ mol Si}}{28 \text{ g Si}} \times \frac{6.02 \times 10^{23} \text{ atoms Si}}{1 \text{ mol Si}} = 4.9 \times 10^{18} \text{ atoms Si}$$

Note the two factors used to make the conversion from grams of silicon to atoms. The first factor is the molar mass of silicon. Such values will always be available to you from the periodic table that accompanies all ACS exams. The value of Avogadro's number will also be given in a table of useful information; or, sometimes it is included with the question itself. Since you use this value so often, you probably already know it. These two conversion factors are arranged so that the units cancel in making the change from grams through moles to atoms. If the conversion were desired in the reverse direction, then both factors would be inverted.

---

**ST-4.** A hydrocarbon with general formula $C_xH_y$ was burned completely in air, yielding 0.18 g of water and 0.44 g of carbon dioxide. Which formula could give such data?

| Formula | Molar Mass |
|---|---|
| $H_2O$ | 18.0 g·mol$^{-1}$ |
| $CO_2$ | 44.0 g·mol$^{-1}$ |

(**A**) $C_2H_2$ (**B**) $C_2H_4$ (**C**) $C_2H_6$ (**D**) $C_6H_6$

*Knowledge Required:* (1) The expected products of combustion reactions of hydrocarbons. (2) The interpretation of experimental data. (3) The determination of mole ratio in a formula.

*Thinking it Through:* This is the general equation for the complete combustion of a hydrocarbon.

$$C_xH_y + (x + y/4)O_2 \rightarrow xCO_2 + y/2H_2O$$

What is needed in this problem is the mole ratio of carbon to hydrogen in the hydrocarbon. The given masses of the products of complete combustion, water, and carbon dioxide, can be used to determine the needed mole ratio. By inspection, it can be seen that 0.44 g of $CO_2$ ($M = 44.0$ g·mol$^{-1}$) is 0.010 mol of $CO_2$. Similarly, in 0.18 g of $H_2O$ ($M = 18.0$ g·mol$^{-1}$), there are 0.010 mol of $H_2O$.

*Note:* In this problem there has been an attempt to simplify the number for you to save unnecessary calculation. If the shortcut is not clear to you, you can always do the formal calculation.

$$0.44 \text{ g } CO_2 \times \frac{1 \text{ mol } CO_2}{44.0 \text{ g } CO_2} = 0.010 \text{ mol } CO_2 \qquad 0.18 \text{ g } H_2O \times \frac{1 \text{ mol } H_2O}{18.0 \text{ g } H_2O} = 0.010 \text{ mol } H_2O$$

All of the carbon in the product carbon dioxide came from the original hydrocarbon; this means there is 0.010 mol C in $C_xH_y$. All of the hydrogen in the product water also came from the original hydrocarbon. The formula shows that there are two moles of hydrogen atoms in every mole of water molecules, so there are 0.010 mol of H in $C_xH_y$. Putting these observations together, there are twice the number of moles of hydrogen as carbon, fixing the formula as $C_2H_4$ which is choice (**B**).

The most common error in this type of problem is overlooking important subscripts, such as the "2" in the formula for water. If you make this error, however, you should be able to recover in this case, for there are two choices, both $C_2H_2$ and $C_6H_6$, that have a one-to-one mole ratio of carbon to hydrogen. Since there is no further information given that would enable you to differentiate between these two responses, you are alerted to check for an error.

---

| ST-5. | Calculate the mass of $SbF_3$ needed to produce 1.00 g of Freon–12, $CCl_2F_2$. The reaction is represented by this equation.  $$3CCl_4 + 2SbF_3 \rightarrow 3CCl_2F_2 + 2SbCl_3$$ | Formula | Molar Mass |
|---|---|---|---|
| | | $SbF_3$ | 179 g·mol$^{-1}$ |
| | | $CCl_2F_2$ | 121 g·mol$^{-1}$ |

| | (A) | 0.667 g | (B) | 0.986 g | (C) | 1.48 g | (D) | 2.22 g |
|---|---|---|---|---|---|---|---|---|

*Knowledge Required:* (1) The method for carrying out mass to mole conversions. (2) The interpretation of a balanced chemical equation.

*Thinking it Through:* The coefficients in a balanced chemical equation give the relative number of moles of each reactant and product. The first step is to change 1.00 g of $CCl_2F_2$ to moles of $CCl_2F_2$, using the given molar mass. An inspection of the coefficients in the balanced equation reveals that two moles of $SbF_3$ are needed to produce every three moles of $CCl_2F_2$. The last step is to use the molar mass of $SbF_3$ to change the moles of $SbF_3$ to grams of $SbF_3$. Here is the mathematical solution.

$$1.00 \text{ g } CCl_2F_2 \times \frac{1 \text{ mol } CCl_2F_2}{121 \text{ g } CCl_2F_2} \times \frac{2 \text{ mol } SbF_3}{3 \text{ mol } CCl_2F_2} \times \frac{179 \text{ g } SbF_3}{1 \text{ mol } SbF_3} = 0.986 \text{ g } SbF_3$$

The correct answer is in choice (**B**). You should realize that other possible answers have been carefully crafted to detect likely errors in solving this problem. For example, if you neglect the 2:3 mole ratio from the balanced equation, you will incorrectly calculate the mass of $SbF_3$ required to be 1.48 grams, choice (**A**). If you neglect to change grams to moles, but apply the mole ratio correctly, you will calculate the mass of $SbF_3$ required as 0.667 g; this is also wrong but it is choice (**A**). If the mole ratio is inverted in error, you will think 2.22 g is the correct answer, choice (**D**).

**ST-6.** What volume of 0.100 M $SO_3^{2-}$(aq) is needed to titrate 24.0 mL of 0.200 M $Fe^{3+}$(aq)? This equation represents the reaction that takes place during the titration.

$$2Fe^{3+}(aq) + SO_3^{2-}(aq) + H_2O(l) \rightarrow 2Fe^{2+}(aq) + SO_4^{2-}(aq) + 2H^+(aq)$$

**(A)** 48.0 mL **(B)** 24.0 mL **(C)** 12.0 mL **(D)** 6.00 mL

*Knowledge Required:* (1) The definition of molarity, M. (2) The interpretation of a balanced chemical equation.

*Thinking it Through:* Often chemical arithmetic is based not on a mass measurement, but on a volume measurement. The molarity of a solution expresses the number of moles of solute in a liter of solution. It is straightforward in this case to predict the volume of $SO_3^{2-}$(aq) that will be required. The $SO_3^{2-}$(aq) is *half* as concentrated as the $Fe^{3+}$(aq) solution, but only *half* the number of moles are required according to the balanced equation. Therefore, 24.0 mL of the $SO_3^{2-}$(aq) will be required for the titration to be completed. This is choice **(B)**.

You may need the full mathematical solution in other cases where the numbers are not so simple. It is equally correct and often more convenient for laboratory work to interpret molarity as the number of millimoles of solute in a milliliter of solution; the ratio of solute to solution remains constant. The coefficients in the balanced ionic equation give the relative number of moles of each reactant and product; this ratio can also be expressed in millimoles.

$$24.0 \text{ mL Fe}^{3+}(aq) \times \frac{0.200 \text{ mmol Fe}^{3+}(aq)}{1 \text{ mL Fe}^{3+}(aq)} \times \frac{1 \text{ mmol SO}_3^{2-}(aq)}{2 \text{ mmol Fe}^{3+}(aq)} \times \frac{1 \text{ mL SO}_3^{2-}(aq)}{0.100 \text{ mmol SO}_3^{2-}(aq)} = 24.0 \text{ mL SO}_3^{2-}(aq)$$

Note that if you omit the 1:2 ratio of $SO_3^{2-}$(aq) to $2Fe^{3+}$(aq), you will erroneously calculate choice **(A)**. Choices **(C)** and **(D)** do not meet a standard of reasonableness. It cannot possibly take *less* than 24.0 mL of 0.100 M $SO_3^{2-}$(aq) to react with 24.0 mL of the more concentrated $Fe^{3+}$(aq).

---

**ST-7.** If a 17.0 g sample of impure nickel metal reacts with excess carbon monoxide, CO, forming 6.25 L of $Ni(CO)_4$ gas under standard temperature and pressure conditions, what is the percent by mass of nickel in the impure nickel metal sample?

$$Ni(s) + 4CO(g) \rightarrow Ni(CO)_4(g)$$

| Formula | Molar Mass |
|---|---|
| Ni | 58.7 g·mol$^{-1}$ |
| Ni(CO)$_4$ | 171 g·mol$^{-1}$ |

**(A)** 24.1% **(B)** 25.0% **(C)** 96.3% **(D)** 100%

*Knowledge Required:* (1) The relationship between the volume of a gas at standard conditions and the number of moles present. (2) The interpretation of a balanced chemical equation. (3) The criteria for distinguishing essential information from nonessential information.

*Thinking it Through:* It is not a common practice on ACS exams, but do not assume that all information provided in a question is actually essential to its solution. Rather, decide on a route to the solution that will be the most expedient. Given that 6.25 L of $Ni(CO)_4$(g) form at standard temperature and pressure conditions, the molar volume of 22.4 L can be used as a conversion factor to find the number of moles of $Ni(CO)_4$ present. Then the coefficients in the balanced chemical equation can be applied, observing that one mole of $Ni(CO)_4$ reacts to produce one mole of Ni. The atomic molar mass for nickel can then be used to find the number of grams of Ni present.

$$6.25 \text{ L Ni(CO)}_4 \times \frac{1 \text{ mol Ni(CO)}_4}{22.4 \text{ L Ni(CO)}_4 \text{ at STP}} \times \frac{1 \text{ mol Ni}}{1 \text{ mol Ni(CO)}_4} \times \frac{58.7 \text{ g Ni}}{1 \text{ mol Ni}} = 16.4 \text{ g Ni}$$

The final step is to find the percent by mass of nickel in the original sample.

$$\% \text{ Nickel} = \frac{16.4 \text{ g pure Ni}}{17.0 \text{ g impure Ni metal sample}} \times 100 = 96.3\% \text{ Ni}$$

Note you do not need to use the molar mass of $Ni(CO)_4$. Because it is a gas at standard temperature and pressure conditions, the molar volume is a useful conversion factor to change volume to moles.

**ST-8.** The combustion reaction of $C_3H_8O$ with $O_2$ is represented by this balanced chemical equation.

$$2C_3H_8O + 9O_2 \rightarrow 6CO_2 + 8H_2O$$

| Formula | Molar Mass |
|---------|-----------|
| $C_3H_8O$ | 60.1 g·mol$^{-1}$ |
| $O_2$ | 32.0 g·mol$^{-1}$ |

When 3.00 g $C_3H_8O$ and 7.38 g $O_2$ are combined, how many moles of which reagent remain?

(A)  0.006 mol $O_2$

(B)  0.024 mol $C_3H_8O$

(C)  0.24 mol $O_2$

(D)  0.18 mol $C_3H_8O$

---

*Knowledge Required:* (1) The identification of the limiting reagent. (2) The method for carrying out mass to mole conversions. (3) The interpretation of a balanced chemical equation.

*Thinking it Through:* Stoichiometry problems often give an exact mass of one reactant and then either explicitly state or implicitly assume that the other reactant is present in excess. The reactants given in this problem are *not* assumed to be present in stoichiometric amounts; this identifies the question as one involving a limiting reagent. The balanced chemical equation shows that $O_2$ with $C_3H_8O$ will react in a 9:2 mole ratio. The first task becomes identification of which material will be completely used up, and therefore which one is present in excess. Although there are several approaches to finding the limiting reagent, they all involve the conversion of grams to moles so that the mole ratios can be compared.

$$7.38 \text{ g } O_2 \times \frac{1 \text{ mol } O_2}{32.0 \text{ g } O_2} = 0.231 \text{ mol } O_2 \quad \text{and} \quad 3.00 \text{ g } C_3H_8O \times \frac{1 \text{ mol } C_3H_8O}{60.1 \text{ g } C_3H_8O} = 0.0499 \text{ mol } C_3H_8O$$

Now these ratios must be compared to the 9:2 mol ratio given in the balanced equation. This can either be done by inspection, or if the ratio is quite different from the stoichiometric ratio, by actual calculation. By inspection, note that there are approximately 0.23 mol of $O_2$ and 0.05 mol of $C_3H_8O$. Is this greater or lower than a 9:2 ratio? This is a good time to actually calculate the ratio, for the ratio will be close to 9:2.

$$\frac{0.231 \text{ mol } O_2}{0.0499 \text{ mol } C_3H_8O} = \frac{4.63 \text{ mol } O_2}{1 \text{ mol } C_3H_8O} \text{ or } \frac{9.26 \text{ mol } O_2}{2 \text{ mol } C_3H_8O}$$

Now it is clear that there is more than enough $O_2$ to react with the $C_3H_8O$ that is present. This identifies $C_3H_8O$ as the limiting reagent, and means that some oxygen will be left over, eliminating choices (B) and (D) from consideration. There will not be a great deal of $O_2$ left over, which makes choice (C) seem unlikely. To be sure, you can calculate the amount of $O_2$ that will exactly react with the 0.0499 mol $C_3H_8O$.

$$0.0499 \text{ mol } C_3H_8O \times \frac{9 \text{ mol } O_2}{2 \text{ mol } C_3H_8O} = 0.225 \text{ mol } O_2 \text{ required for a complete reaction with all available } C_3H_8O$$

The difference between the available moles of $O_2$ (0.231 mol) and the moles of $O_2$ required for a complete reaction with the available $C_3H_8O$ (0.225 mol) is what will remain after the reaction goes to completion. Subtracting gives 0.006 mol, confirming choice (A) as the correct response.

---

**ST-9.** What is the maximum mass of $(NH_4)_2SO_4$ that could be formed from 17 kg of $NH_3$ and 200 kg of solution containing 49% $H_2SO_4$ by mass? This equation represents the reaction.

$$2NH_{3(g)} + H_2SO_{4(aq)} \rightarrow (NH_4)_2SO_{4(s)}$$

| Formula | Molar Mass |
|---------|-----------|
| $NH_3$ | 17 g·mol$^{-1}$ |
| $H_2SO_4$ | 98 g·mol$^{-1}$ |
| $(NH_4)_2SO_4$ | 132 g·mol$^{-1}$ |

(A)  217 kg    (B)  132 kg    (C)  115 kg    (D)  66 kg

*Knowledge Required:* (1) The identification of the limiting reagent. (2) The method for carrying out mass–to–mole conversions. (3) The interpretation of a balanced chemical equation.

*Thinking it Through:* There are two reactants given, so the first step is to determine if there is a limiting reagent. There is no need to change kilograms to grams. The molar mass can be interpreted either as the number of grams per mole or the number of kilograms per kilomole. Compare the number of kilomoles of each reactant, either by inspection or by calculation.

$$17 \text{ kg NH}_3 \times \frac{1 \text{ kmol NH}_3}{17 \text{ kg NH}_3} = 1.0 \text{ kmol NH}_3 \text{ and}$$

$$200 \text{ kg H}_2\text{SO}_4 \text{ solution} \times \frac{49 \text{ kg H}_2\text{SO}_4}{100 \text{ kg H}_2\text{SO}_4 \text{ solution}} \times \frac{1 \text{ kmol H}_2\text{SO}_4}{98 \text{ kg H}_2\text{SO}_4} = 1.0 \text{ kmol H}_2\text{SO}_4$$

The chemical equation shows that there is a 2:1 mol ratio, $NH_3$ to $H_2SO_4$. Therefore for 1.0 kmol of $NH_3$, only 0.5 kmol of $H_2SO_4$ is required, making it the reactant present in excess. $NH_3$ is the limiting reagent.

$$1.0 \text{ kmol NH}_3 \times \frac{1 \text{ kmol (NH}_4\text{)}_2\text{SO}_4}{2 \text{ kmol NH}_3} \times \frac{132 \text{ kg (NH}_4\text{)}_2\text{SO}_4}{1 \text{ kmol (NH}_4\text{)}_2\text{SO}_4} = 66 \text{ kg (NH}_4\text{)}_2\text{SO}_4$$

This is choice (**D**). If you incorrectly choose the sulfuric acid as the limiting reagent, you will obtain 132 kg, which is choice (**B**). Choice (**A**) is just the direct sum of the two given masses; not a correct approach. Choice (**C**) might be obtained by adding the mass of both $NH_3$ and pure $H_2SO_4$, an incorrect use of the Law of Conservation of Mass.

---

**ST-10.** What is the percent yield of $PbI_2$ if 5.00 g of $PbI_2$ results from a solution containing 10.0 g of $Pb(C_2H_3O_2)_2$ with a solution containing an excess of KI? This is the equation representing the reaction. $Pb(C_2H_3O_2)_2(aq) + 2KI(aq) \rightarrow PbI_2(s) + 2KC_2H_3O_2(aq)$

| Formula | Molar Mass |
|---|---|
| $PbI_2$ | 461.0 g·mol$^{-1}$ |
| $Pb(C_2H_3O_2)_2$ | 325.3 g·mol$^{-1}$ |

(**A**)    17.6%       (**B**)    35.3%       (**C**)    50.0%       (**D**)    70.5%

---

*Knowledge Required:* (1) The meaning of percent yield. (2) The method for carrying out mass–to–mole conversions. (3) The interpretation of a balanced chemical equation.

*Thinking it Through:* The percent yield of a chemical reaction is a ratio of the experimental yield to the maximum amount that could theoretically be produced. This is the relationship.

$$\text{Percent Yield} = \frac{\text{Experimental Yield}}{\text{Theoretical Yield}} \times 100$$

In this problem, the experimental yield of 5.00 g $PbI_2$ is given. Before finding the percent yield, it is necessary to determine the theoretical yield, which is the maximum amount of product that could be produced. The theoretical yield is based on the given amounts of reactants, taking into consideration the mole ratios of reactants to products revealed in the balanced chemical equation. There is an excess of KI solution. The 10.0 g of $Pb(C_2H_3O_2)_2$ is the limiting reagent and determines the maximum amount of $PbI_2(s)$ that could be produced. Observe that there is a one-to-one mole ratio between the reactant $Pb(C_2H_3O_2)_2$ and the product $PbI_2(s)$.

$$10.0 \text{ g Pb(C}_2\text{H}_3\text{O}_2\text{)}_2 \times \frac{1 \text{ mol Pb(C}_2\text{H}_3\text{O}_2\text{)}_2}{325.3 \text{ g Pb(C}_2\text{H}_3\text{O}_2\text{)}_2} \times \frac{1 \text{ mol PbI}_2}{1 \text{ mol Pb(C}_2\text{H}_3\text{O}_2\text{)}_2} \times \frac{461.0 \text{ g PbI}_2}{1 \text{ mol PbI}_2} = 14.2 \text{ g PbI}_2$$

Now there is enough information to compare the experimental yield of 5.00 g to the theoretical yield of 14.2 g.

$$\text{Percent Yield} = \frac{\text{Experimental Yield}}{\text{Theoretical Yield}} \times 100 = \frac{5.00 \text{ g PbI}_2}{14.2 \text{ g PbI}_2} \times 100 = 35.3 \text{ percent yield}$$

*Note:* In this case, the balanced equation is given. In other cases, you may have to provide the equation or balance a given skeleton equation before solving the problem.

## Practice Questions

1. A compound is found to consist of 34.9% sodium, 16.4% boron and 48.6% oxygen. What is its simplest formula?

   (A)  $NaBO_2$     (B)  $NaBO_3$     (C)  $Na_2B_4O_7$     (D)  $Na_3BO_3$

2. Upon analysis, a compound is found to contain 22.8% sodium, 21.8% boron, and 55.4% oxygen. What is its simplest formula?

   (A)  $NaBO$     (B)  $NaB_2O_5$     (C)  $Na_2B_4O_7$     (D)  $Na_3BO_4$

3. A 4.08 g sample of a compound of nitrogen and oxygen contains 3.02 g of oxygen. What is the empirical formula?

   (A)  $NO$     (B)  $NO_2$     (C)  $N_2O$     (D)  $N_2O_5$

4. What is the percent by mass of oxygen in $Ca(NO_3)_2$?

   (A)  29.3 %     (B)  47.1 %     (C)  58.5 %     (D)  94.1 %

5. Which of these compounds contains the greatest percentage of nitrogen?

   | Formula | Molar Mass |
   | --- | --- |
   | $C_6H_3N_3O_7$ | 229. g·mol$^{-1}$ |
   | $CH_4N_2O$ | 60.1 g·mol$^{-1}$ |
   | $LiNH_2$ | 23.0 g·mol$^{-1}$ |
   | $Pb(N_3)_2$ | 291. g·mol$^{-1}$ |

   (A)  $C_6H_3N_3O_7$     (B)  $CH_4N_2O$     (C)  $LiNH_2$     (D)  $Pb(N_3)_2$

6. What mass of carbon is present in 0.500 mol of sucrose ($C_{12}H_{22}O_{11}$)?

   | Formula | Molar Mass |
   | --- | --- |
   | $C_{12}H_{22}O_{11}$ | 342 g·mol$^{-1}$ |

   (A)  60.0 g     (B)  72.0 g     (C)  90.0 g     (D)  120. g

7. How many atoms are in 1.50 g of Al?

   (A)  0.0556     (B)  18.0     (C)  $3.35 \times 10^{22}$     (D)  $2.44 \times 10^{25}$

8. A sample of a compound of xenon and fluorine contains molecules of a single type; $XeF_n$, where $n$ is a whole number. If $9.03 \times 10^{20}$ of these $XeF_n$ molecules have a mass of 0.311 g, what is the value of $n$?

   (A)  2     (B)  3     (C)  4     (D)  6

9. What mass of carbon is present in $1.4 \times 10^{20}$ molecules of sucrose ($C_{12}H_{22}O_{11}$)?

   | Formula | Molar Mass |
   | --- | --- |
   | $C_{12}H_{22}O_{11}$ | 342 g·mol$^{-1}$ |

   (A)  $1.7 \times 10^{21}$ g     (B)  $2.0 \times 10^{22}$ g     (C)  $3.3 \times 10^{-2}$ g     (D)  $2.8 \times 10^{-3}$ g

10. A 2.000 g sample of an unknown metal, M, was completely burned in excess $O_2$ to yield 0.02224 mol of the metal oxide, $M_2O_3$. What is the metal?

    (A)  Y     (B)  Ca     (C)  Al     (D)  Sc

11. A $3.41 \times 10^{-6}$ g sample of a compound is known to contain $4.67 \times 10^{16}$ molecules. This compound is

    (A)  $CO_2$     (B)  $CH_4$     (C)  $NH_3$     (D)  $H_2O$

12. Avogadro's number equals the number of

    (A)  atoms in one mole of atoms.     (B)  molecules in one mole of $O_2$.

    (C)  marbles in one mole of marbles.     (D)  all of the above.

13. The number of atoms in 9.0 g of aluminum is the same as the number of atoms in

    (A)  8.1 g of magnesium.     (B)  9.0 g of magnesium.

    (C)  12.1 g of magnesium.     (D)  18.0 g of magnesium.

14. A single molecule of a certain compound has a mass of $3.4 \times 10^{-22}$ g. Which value comes closest to the mass of a mole of this compound?

    (A)  $50$ g·mol$^{-1}$     (B)  $100$ g·mol$^{-11}$     (C)  $150$ g·mol$^{-1}$     (D)  $200$ g·mol$^{-1}$

15. What mass of $KClO_3$ will produce 48.0 g of oxygen gas, $O_2$, if the decomposition of $KClO_3$ is complete?

| Formula | Molar Mass |
|---|---|
| $KClO_3$ | 122.5 g·mol$^{-1}$ |

    (A)  61.3 g     (B)  74.5 g     (C)  122.5 g     (D)  245.0 g

16. What is the maximum mass of aluminum chloride that could be obtained from 6.00 mol of barium chloride and excess aluminum sulfate? This is the balanced equation for the reaction.
    $$Al_2(SO_4)_3 + 3BaCl_2 \rightarrow 3BaSO_4 + 2AlCl_3$$

| Formula | Molar Mass |
|---|---|
| $BaCl_2$ | 208.3 g·mol$^{-1}$ |
| $AlCl_3$ | 133.3 g·mol$^{-1}$ |

    (A)  1250 g     (B)  801 g     (C)  534 g     (D)  134 g

17. A self-contained breathing apparatus uses potassium superoxide, $KO_2$, to convert the carbon dioxide and water in exhaled air into oxygen, as shown by this equation.
    $$4KO_{2(s)} + 4CO_{2(g)} + 2H_2O_{(g)} \rightarrow 4KHCO_{3(s)} + 3O_{2(g)}$$
    How many molecules of oxygen gas will be produced from the 0.0468 g of carbon dioxide that is exhaled in a typical breath?

    (A)  $4.8 \times 10^{20}$     (B)  $6.4 \times 10^{20}$     (C)  $8.5 \times 10^{20}$     (D)  $1.9 \times 10^{21}$

18. A mixture containing 9 mol of $F_2$ and 4 mol of S is allowed to react. This equation represents the reaction that takes place.
    $$3F_2 + S \rightarrow SF_6$$
    How many moles of $F_2$ *remain* after 3 mol of S have reacted?

    (A)  4     (B)  3     (C)  1     (D)  0

19. When 1.187 g of a metallic oxide is reduced with excess hydrogen, 1.054 g of the metal is produced. What is the metallic oxide?

    (A)  $Ag_2O$     (B)  $Cu_2O$     (C)  $K_2O$     (D)  $Tl_2O$

20. How many moles of iron react with 1.75 mol of oxygen gas? The equation for the reaction is:
    $$3O_{2(g)} + 4Fe_{(s)} \rightarrow 2Fe_2O_{3(s)}$$

    (A)  1.31 mol     (B)  1.75 mol     (C)  2.33 mol     (D)  5.25 mol

21. The limiting reagent in a particular reaction can be recognized because it is the reagent that

    (A)    has the smallest coefficient in the balanced equation.

    (B)    has the smallest mass in the reaction mixture.

    (C)    is present in the smallest molar quantity.

    (D)    would be used up first.

22. Consider this reaction used for the production of lead.

    $$2PbO(s) + PbS(s) \rightarrow 3Pb(s) + SO_2(g)$$

    What is the maximum mass of lead that can be obtained by the reaction of 57.33 g PbO and 33.80 g of PbS?

    | Formula | Molar Mass |
    |---------|------------|
    | Pb | 207.2 g·mol$^{-1}$ |
    | PbO | 223.2 g·mol$^{-1}$ |
    | PbS | 239.3 g·mol$^{-1}$ |
    | SO$_2$ | 64.07 g·mol$^{-1}$ |

    (A)    43.48 g        (B)    72.75 g        (C)    79.83 g        (D)    87.80 g

23. What volume of 0.131 M $BaCl_2$ is required to react completely with 42.0 mL of 0.453 M $Na_2SO_4$? This is the net ionic equation for the reaction.

    $$Ba^{2+}(aq) + SO_4^{2-}(aq) \rightarrow BaSO_4(s)$$

    (A)    12.1 mL        (B)    72.6 mL        (C)    145 mL        (D)    290 mL

24. In acidic solution, the dichromate ion, $Cr_2O_7^{2-}(aq)$ will oxidize $Fe^{2+}$ to $Fe^{3+}$ and form $Cr^{3+}$. This net ionic equation represents the reaction that takes place during the reaction.

    $$Cr_2O_7^{2-}(aq) + 6Fe^{2+}(aq) + 14H^+(aq) \rightarrow 2Cr^{3+}(aq) + 6Fe^{3+}(aq) + 7H_2O(l)$$

    What volume of 0.100 M $Cr_2O_7^{2-}(aq)$ is required to oxidize 60.0 mL of 0.250 M $Fe^{2+}(aq)$?

    (A)    25.0 mL        (B)    42.0 mL        (C)    58.4 mL        (D)    175 mL

25. Consider this reaction. $CaCO_3(s) \rightarrow CaO(s) + CO_2(g)$

    What mass of $CaCO_3$ will produce 8.0 L of $CO_2$, measured at standard temperature and pressure conditions?

    | Formula | Molar Mass |
    |---------|------------|
    | CaCO$_3$ | 100. g·mol$^{-1}$ |

    (A)    4.5 g        (B)    12.5 g        (C)    36 g        (D)    280 g

26. When $FeCl_3$ is ignited in an atmosphere of pure oxygen, this reaction takes place.

    $$4FeCl_3(s) + 3O_2(g) \rightarrow 2Fe_2O_3(s) + 6Cl_2(g)$$

    If 3.00 mol of $FeCl_3$ are ignited in the presence of 2.00 mol of $O_2$ gas, how much of which reagent is present in excess and therefore remains unreacted?

    (A)    0.33 mol $FeCl_3$ remains unreacted        (B)    0.67 mol $FeCl_3$ remains unreacted

    (C)    0.25 mol $O_2$ remains unreacted        (D)    0.50 mol $O_2$ remains unreacted

27. Ammonia, $NH_3$, reacts with the hypochlorite ion, $OCl^-$, to produce hydrazine, $N_2H_4$. How many moles of hydrazine are produced from 5.85 mol of ammonia if the reaction has a 78.2% yield?

    $$2NH_3 + OCl^- \rightarrow N_2H_4 + Cl^- + H_2O$$

    (A)    2.29 mol        (B)    2.92 mol        (C)    4.57 mol        (D)    9.15 mol

**28.** Ammonia can be prepared by the Haber process, shown in this equation.

$$N_2 + 3H_2 \rightleftharpoons 2NH_3$$

If 2 mol of $N_2$ and 3 mol of $H_2$ are combined, the amount of $NH_3$ that would be formed if all of the limiting reactant were used up is known as the

(A) limited yield. (B) percent yield. (C) product yield. (D) theoretical yield.

**29.** Antimony reacts with sulfur according to this equation.

| Formula | Molar Mass |
|---------|------------|
| $Sb_2S_3$ | 339.7 g·mol$^{-1}$ |

$$2Sb(s) + 3S(s) \rightarrow Sb_2S_3(s)$$

What is the percentage yield for a reaction in which 1.40 g of $Sb_2S_3$ is obtained from 1.73 g of antimony and a slight excess of sulfur?

(A) 80.9% (B) 58.0% (C) 40.5% (D) 29.0%

**30.** When 4.50 g of $Fe_2O_3$ is reduced with excess $H_2$ in a furnace, 2.60 g of metallic iron is recovered. What is the percent yield? This is the equation representing the reaction.

| Formula Molar Mass | |
|--------------------|--|
| $Fe_2O_3$ | 159.7 g·mol$^{-1}$ |

$$Fe_2O_3(s) + 3H_2(g) \rightarrow 2Fe(s) + 3H_2O(g)$$

(A) 82.6% (B) 70.0% (C) 57.8% (D) 31.5%

## Answers to Study Questions

| | | | | | |
|---|---|---|---|---|---|
| 1. | A | 5. | B | 9. | D |
| 2. | D | 6. | B | 10. | B |
| 3. | A | 7. | C | | |
| 4. | B | 8. | A | | |

## Answers to Practice Questions

| | | | | | |
|---|---|---|---|---|---|
| 1. | A | 11. | A | 21. | D |
| 2. | C | 12. | D | 22. | C |
| 3. | D | 13. | A | 23. | C |
| 4. | C | 14. | D | 24. | A |
| 5. | C | 15. | C | 25. | C |
| 6. | B | 16. | C | 26. | A |
| 7. | C | 17. | A | 27. | A |
| 8. | C | 18. | D | 28. | D |
| 9. | C | 19. | B | 29. | B |
| 10. | D | 20. | C | 30. | A |

# States of Matter / Solutions

The three common states of matter–solids, liquids, and gases–are considered in most general chemistry courses. Both qualitative generalizations and quantitative laws are part of this study. Phase diagrams show the temperature/pressure relationships among the three states. The relative physical properties of the three states are studied, including diffusion and compressibility. For solids, other topics may include bonding, crystal shapes, and solubility rules. For liquids, properties such as miscibility and vapor pressure are included. Properties of gases and mathematical relationships such as the ideal gas laws are either explicitly taught or assumed as a common base of knowledge. Topics considered with solutions are units of concentration, preparation and dilution, and properties associated with the number of solute particles dissolved.

## Study Questions

**SM-1.** According to this phase diagram showing the gas, liquid, and solid phases of a pure substance, what phase or phases can be present at point **X**?

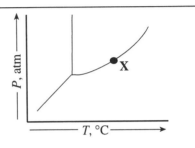

| | | | |
|---|---|---|---|
| **(A)** | liquid only | **(B)** | liquid and gas only |
| **(C)** | solid and gas only | **(D)** | liquid and solid only |

*Required Knowledge:* (1) The states of matter. (2) The interpretation of phase diagrams.

*Thinking it Through:* A generalized phase diagram shows how the solid, liquid, and gaseous regions vary with temperature and pressure. The three phases are always in the same relative positions; the slopes of the lines vary considerably. The lines show the boundaries between phases and under those conditions, both phases occur together. Point **X** is located on the phase boundary between liquid and gaseous states, so choice **(B)** is correct.

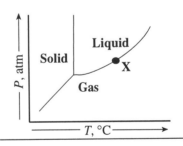

**SM-2.** What is the normal melting point of the pure substance **Z**?

**Phase Diagram for Substance Z**

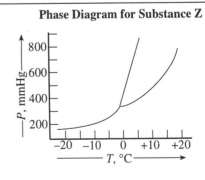

| | | | | | | | |
|---|---|---|---|---|---|---|---|
| **(A)** | +18 °C | **(B)** | +4 °C | **(C)** | –2 °C | **(D)** | –22 °C |

*Required Knowledge:* (1) States of matter. (2) Interpretation of phase diagrams. (3) Definition of melting point.

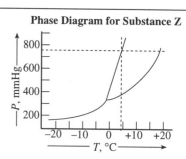

*Thinking it Through:* The normal melting point of a substance is the temperature at which the solid and liquid phases are in equilibrium at standard atmospheric pressure, 760 mmHg. From the graph, the temperature corresponding to 760 mmHg is approximately +4 °C, making choice (**B**), the correct answer. Choice (**A**) gives the approximate boiling point, not the melting point. Choice (**C**) is the triple point for this substance, the temperature and pressure at which all three states occur in equilibrium. Choice (**D**) has no significance but just appears to be the left end of the curve.

---

**SM-3.**    Which substance meets the requirements: cubic solid; poor electrical conductor in the solid state; good conductor when fused?

(**A**)    sodium chloride                  (**B**)    diamond

(**C**)    sulfur                            (**D**)    chromium

---

*Knowledge Required:* (1) The bonding in solids. (2) The properties associated with bonding types.

*Thinking it Through:* Solids exhibit a variety of bonding types, each of which can be associated with different properties. This table summarizes the expected relationships.

| Type of bonding in solid | Structure of solid | Expected Conductivity of solid | Expected Conductivity if solid is fused |
|---|---|---|---|
| Ionic | crystalline | low | high |
| Covalent | amorphous or crystalline | low | low |
| Network covalent | crystalline or amorphous | low | low |
| Metallic | crystalline | high | high |

Of the solids given, ionically bonded sodium chloride is expected to be crystalline, a poor electrical conductor in the solid form, and a good conductor when fused. Diamond, formed of covalently bonded carbon atoms, is a network substance that does not form cubic crystalline patterns, and does not conduct electricity either when solid or fused. None of the allotropic forms of sulfur is expected to conduct electricity. Choice (**D**), the metal chromium, could possibly form a cubic solid crystalline form, but can be eliminated because it is expected to conduct electricity both when a solid and when fused. The correct choice is (**A**), because sodium chloride is a crystalline solid that is a poor conductor in the solid state and a good conductor when fused.

---

**SM-4.**    Two pure solid substances, **A** and **B**, each melt at 87.5 °C. An unknown pure solid substance **X** also melts at 87.5 °C. Melting points are obtained for mixtures of **X** with **A**, **X** with **B**, and **A** with **B**. The melting point of each mixture tested is found to be below 87.5 °C. Which conclusion is justified by these data?

(**A**)    **X** is the same as **A** but different from **B**.

(**B**)    **X**, **A**, and **B** are all the same substance.

(**C**)    **X** is different from both **A** and **B**.

(**D**)    **A** and **B** are the same, but **X** is different.

---

*Knowledge Required:* (1) The relationship of melting point with purity. (2) The interpretation of experimental data.

*Thinking it Through:* Pure substances can be differentiated based on the temperature at which the solid and liquid phases are in equilibrium. This temperature is termed the melting point. Each of the pure substances **A**, **B**, and **X** has the *same* melting point. If these were different samples of the same substance, then any mixture of **A**, **B**, and **X** would also have the *same* melting point. This is not consistent with the given observations, so choice (**B**) can be eliminated. So can choice (**D**), because the question states that substance **A** and **B** are different. If different substances are mixed, even if they have the same melting point when pure, the melting point of the mixture is expected to be *lower* than the melting point of the pure substance. None of the three combinations produces a melting point of 87.5 °C, so all three substances *must* be different. This eliminates choice (**A**) and again eliminates choice (**D**). This leaves choice (**C**), in which **X** is different from both substance **A** and substance **B**.

---

**SM-5.** The vapor pressure of four different substances at 20 °C is given in the table. Which statement about these substances is correct?

| Substance | Vapor pressure, mmHg |
|---|---|
| 1 | 0.0012 |
| 2 | 18 |
| 3 | 175 |
| 4 | 442 |

(A) The normal boiling point of **4** should be greater than that of **2**.

(B) The heat of vaporization of **3** should be greater than that of **2**.

(C) The intermolecular forces between molecules of **2** should be greater than those between molecules of **1**.

(D) The surface tension of **2** should be greater than that of **3**.

*Knowledge Required:* (1) The meaning of term vapor pressure. (2) The correlation of vapor pressure data with other physical properties.

*Thinking it Through:* Substances with a lower numerical value of vapor pressure have little tendency to change from liquid to gas. Such substances have strong particle-to-particle interactions and/or high average molecular mass. These substances can be expected to have relatively higher heats of vaporization, higher boiling points, and greater surface tension. Larger values of vapor pressure indicate substances with low average molecular mass and/or with very little interaction among the particles. These substances will have relatively lower heats of vaporization, lower boiling points, and less surface tension. Choice (**A**) is not reasonable based on the given trend in vapor pressure, because substance **4** has a higher vapor pressure than substance **2**, indicating it will have a lower boiling point. Choices (**B**) and (**C**) are also reversed from the expected trend. Choice (**D**) is the most reasonable. Substance **2** with its lower vapor pressure than substance **3**, is expected to have a greater surface tension.

---

**SM-6.** A mixture of 0.5 mol of $CH_4$, 0.5 mol of $H_2$, and 0.5 mol of $SO_2$ is introduced into a 10.0-L container at 25 °C. If the container has a pinhole leak, which describes the relationship between the partial pressures of the individual components in the container after 3 hours?

(A) $P_{SO_2} > P_{CH_4} > P_{H_2}$

(B) $P_{SO_2} < P_{CH_4} < P_{H_2}$

(C) $P_{SO_2} < P_{CH_4} > P_{H_2}$

(D) $P_{SO_2} = P_{CH_4} = P_{H_2}$

*Required Knowledge:* (1) The behavior of gases. (2) The kinetic–molecular theory. (3) The qualitative understanding of Graham's Law of Effusion.

*Thinking it Through:* Equal amounts of each gas are present, so the partial pressure of each gas remaining in the container after 3 hours will depend on the rate with which each escapes. The lightest gases will escape most quickly through a pinhole leak. The molar mass of $CH_4$ is 18 g·mol$^{-1}$, of $H_2$ is 2 g·mol$^{-1}$, and of $SO_2$ is 64 g·mol$^{-1}$. Hydrogen will escape most quickly, followed by methane, and then sulfur dioxide. The partial pressure of the each gas remaining in the container will therefore be in reverse order, $P_{SO_2} > P_{CH_4} > P_{H_2}$. This is choice (**A**).

**SM-7.** Which is true about equal volumes of $CH_4$ and $O_2$ gases at 20 °C and 1 atm pressure?

**(A)** The $CH_4$ sample has a mass that is one-half that of the $O_2$ sample.

**(B)** The number of $O_2$ molecules is twice as large as the number of $CH_4$ molecules.

**(C)** The average kinetic energy of the $O_2$ molecules is one-half that of the $CH_4$ molecules.

**(D)** The average velocity of the $O_2$ molecules is one-half that of the $CH_4$ molecules

*Knowledge Required:* (1) The understanding of Avogadro's Principle. 2) The relationships of average molecular speed and kinetic energy to temperature and pressure conditions.

*Thinking it Through:* The two gases, $CH_4$ and $O_2$, are at the same temperature and pressure conditions. Avogadro's Principle states that equal volumes of any gas under the same temperature and pressure conditions will contain the same number of molecular particles. Although this straightforward answer is not given as one of the choices, it does eliminate choice **(B)**. If the molar masses of the two gases are considered, 16 g·mol⁻¹ for $CH_4$ and 32 g·mol⁻¹ for $O_2$, the 1:2 ratio is revealed. This means that even though an equal number of particles are present, only half the mass of $CH_4$ is present, making choice **(A)** the correct answer. Choice **(C)** is incorrect as the average kinetic energy is directly proportional to the temperature, which is the same for both gases. Choice **(D)** is also incorrect for the average velocity is inversely proportional to the square root of the mass of the molecules, an application of Graham's Law.

**SM-8.** Which microscopic representation best represents a solution?

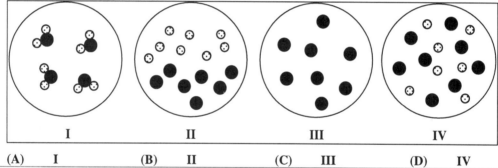

| I | II | III | IV |
|---|---|---|---|

**(A)** I **(B)** II **(C)** III **(D)** IV

*Required Knowledge:* (1) The definition of a solution. (2) The particulate representation of particles forming solutions.

*Thinking it Through:* Solutions are homogeneous mixtures of two or more substances. They may exist in all three phases, although often in general chemistry solutions are aqueous solutions meaning that the solvent is water. Choice **(A)** cannot represent a solution, as there is only one type of molecular particle represented. Choice **(B)** has two different types of particles, but they appear to be in distinct layers. Choice **(C)** again has only one type of particle. Choice **(D)** is the best representation as it shows two types of particles distributed relatively homogeneously.

**SM-9.** A solution is made by dissolving 60 g of NaOH ($M = 40$ g·mol⁻¹) in enough distilled water to make 300 mL of a stock solution. What volumes of this solution and distilled water, when mixed, will result in a solution that is approximately 1 M NaOH?

| | mL stock solution | mL distilled water |
|---|---|---|
| **(A)** | 20 | 80 |
| **(B)** | 20 | 100 |
| **(C)** | 60 | 30 |
| **(D)** | 60 | 90 |

*Knowledge Required:* (1) The definition of molarity. (2) The knowledge of vocabulary associated with solutions. (3) The procedures for preparing dilute solutions.

*Thinking it Through:* Molarity is commonly used to express the concentration of a solution.

$$M = \frac{\text{moles of solute}}{\text{L of solution}} = \frac{\text{grams of solute}}{\text{molar mass of solute} \times \text{L of solution}}$$

The molarity of the solution first prepared can be calculated by substituting the given values. The volume of solution, 300 mL, should be expressed in liters, 0.300 L.

$$M = \frac{\left[60 \text{ g NaOH} \times \left(\dfrac{1 \text{ mol NaOH}}{40 \text{ g}}\right)\right]}{0.300 \text{ L}} = \frac{5 \text{ mol}}{\text{L}} = 5 \text{ M}$$

This solution, called the stock solution, will now be used to prepare a solution that is approximately 1 M NaOH. The number of moles of NaOH in the diluted solution will be the same as it was in the stock solution; it will just be dispersed in a larger volume. Since we are seeking a volume in milliliters, it is more convenient to express the molarity as millimoles per milliliter (which is exactly the same as moles per liter).

$$\text{millimoles NaOH in stock solution} = \text{millimoles of NaOH in diluted solution}$$

Assuming 20 mL of stock solution,

$$(20 \text{ mL})_{\text{stock}} \times (5 \text{ mmol/mL})_{\text{stock}} = V_{\text{new solution}} \times (1 \text{ mmol/mL})_{\text{new solution}}$$

$$V_{\text{new solution}} = \left(\frac{20 \text{ mL} \times 5 \text{ mmol/mL}}{1 \text{ mmol/mL}}\right) = 100 \text{ mL}$$

The total volume of the new solution is approximately equal to the volume of stock solution (20 mL) plus the water added (*x* mL) to produce 100 mL. That comes out to 80 mL of water added, which is intended to be the correct choice.

$$x \text{ mL} = 100 \text{ mL} - 20 \text{ mL} = 80 \text{ mL}$$

You've probably noticed that we use the word *approximately* when describing the concentration and volume of the new solution. In fact, adding 80 mL of water to 20 mL of stock solution produces slightly less than 100 mL of solution. The concentration of NaOH in the diluted solution is slightly greater than 1 M. This is because the densities of NaOH solutions are greater than 1 g/mL. As the NaOH solution is diluted, the density of the solution approaches that of water, which is 1 g/mL. Nonetheless, 80 mL, choice **(A),** is still the best. This is a 1:5 dilution. Choice **(B)** gives a 1:6 dilution, but you may be drawn to it if you neglect to consider that it is not just the water added, but the *total volume* of the solution that must be used. Choice **(C)** results in a 60:90 or a 2:3 dilution, and choice **(D)** results in a 2:5 dilution. None of these produce a 1 M solution.

---

**SM-10.** Given the cooling curve of one gram of a pure liquid, the length of the line **AB** depends on

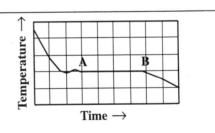

(A) the specific heat of the pure solid.

(B) the specific heat of the pure liquid.

(C) the heat of melting of the pure substance.

(D) the melting point of the pure substance.

*Knowledge Required:* (1)The meaning of specific heat, heat of melting, and melting point. (2) The interpretation of experimental data.

*Thinking it Through:* As a pure liquid cools undisturbed in a clean container, the temperature usually dips briefly below the melting point of the solid. When solid crystals begin to form, the temperature rises to the melting point of the solid (point **A**), where it remains until all the liquid has finally been converted to solid (point **B**). Neither **A** nor **B** are reasonable choices, because both liquid and solid are present from point **A** to point **B**. Choice **(C)** recognizes that the heat being removed from the liquid goes into changing the potential energy of the molecules. Kinetic energy remains the same. Choice **(D)** is an intensive property (temperature) rather than an extensive property (heat). The cooling curve interprets heat being removed from the liquid.

## Practice Questions

1.  In the phase diagram, which transition represents the condensation of a gas to a liquid?

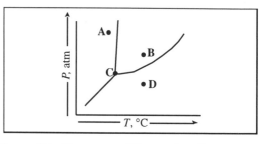

    (A)    B to A        (B)    D to B        (C)    C to D        (D)    A to D

2.  Consider the phase diagram of a pure compound. Which statement applies?

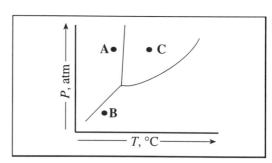

    (A)    The path **A → C** represents sublimation.

    (B)    Following the path **A → B → C,** the compound would first liquefy and then vaporize.

    (C)    If the compound is in state **A**, continued reduction of the pressure (at constant temperature) will cause it to melt.

    (D)    None of these statements is correct.

3.  A particular solid is soft, a poor conductor of heat and electricity, and has a low melting point. Generally, such a solid is classified as

    (A)    ionic.                        (B)    molecular.

    (C)    metallic.                     (D)    network covalent.

4.  Carbon dioxide, $CO_2$, in the form of dry ice would be classified as

    (A)    an ionic solid.               (B)    a molecular solid.

    (C)    a polymeric solid.            (D)    a metallic solid.

5.  The melting point of an impure compound is generally

    (A)    higher than that of the pure solid.

    (B)    the same as that of the pure solid.

    (C)    lower than that of the pure solid.

    (D)    a function of the vapor pressure of the impurity.

6.  What is the vapor pressure of a solution with a benzene to octane molar ratio of 2:1?

    | Vapor Pressure at 50 °C | |
    | --- | --- |
    | benzene | 280 mmHg |
    | octane | 400 mmHg |

    (A)    120 mmHg    (B)    320 mmHg    (C)    400 mmHg    (D)    680 mmHg

7.  Lithium crystallizes in a body-centered cubic unit cell. What is the mass of one unit cell? Report your answer in grams.

    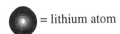 = lithium atom

    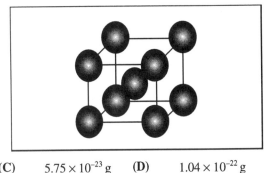

    (A)    $1.15 \times 10^{-23}$ g    (B)    $2.30 \times 10^{-23}$ g    (C)    $5.75 \times 10^{-23}$ g    (D)    $1.04 \times 10^{-22}$ g

8.  Copper crystallizes in a face-centered cubic lattice. If the edge of the unit cell is 362 pm, what is the radius of the copper atom?

     = copper atom

    (A)    128 pm    (B)    171 pm    (C)    181 pm    (D)    255 pm

9.  Which factor affects the vapor pressure of a liquid?

    (A)    temperature                (B)    atmospheric pressure

    (C)    volume of the liquid        (D)    surface area of the liquid

10. A gas or vapor may be liquefied only at temperatures

    (A)    equal to the normal boiling point.

    (B)    above the normal boiling point.

    (C)    above the critical temperature.

    (D)    at or below the critical temperature.

11. The graph shows how the vapor pressure of a liquid changes with temperature. Select the choice that best indicates the degree of correctness of this statement: *"The normal boiling point of the liquid is 78 °C."*

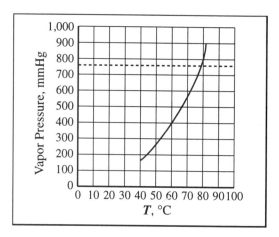

(A)    The statement is true.

(B)    The statement is probably true; additional data would be needed for a final decision.

(C)    The statement is probably false; additional data would be needed for a final decision.

(D)    The statement is false.

12. What is the molar mass of an ideal gas if a 0.622 g sample of this gas occupies a volume of 300. mL at 35 °C and 789 mm Hg?

(A)    44.8 g·mol$^{-1}$    (B)    48.9 g·mol$^{-1}$    (C)    50.5 g·mol$^{-1}$    (D)    54.5 g·mol$^{-1}$

13. Which set of temperature and pressure conditions will cause a gas to exhibit the greatest deviation from ideal gas behavior?

(A)    100 °C and 4 atm              (B)    100 °C and 2 atm

(C)    −100 °C and 4 atm            (D)    0 °C and 2 atm

14. A 500 mL gas sample is collected over water at a pressure of 740 mmHg and 25 °C. What is the volume of the dry gas at STP? (STP = 1 atm and 0 °C.)

| Vapor Pressure of $H_2O$ at 25 °C |
| --- |
| 24 mmHg |

(A)    $500 \times \dfrac{740+24}{760} \times \dfrac{298}{273}$

(B)    $500 \times \dfrac{740-24}{760} \times \dfrac{273}{298}$

(C)    $500 \times \dfrac{760}{740-24} \times \dfrac{273}{298}$

(D)    $500 \times \dfrac{760}{740+24} \times \dfrac{298}{273}$

15. Which diagram represents the most concentrated solution?

Note: ⬤ represents a solute particle and ⊙–⊙ represents a water molecule.

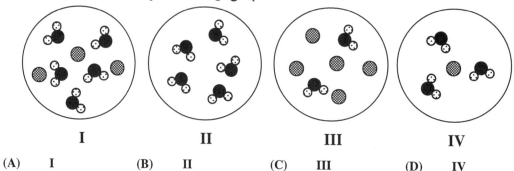

I                     II                     III                     IV

(A)    I              (B)    II             (C)    III            (D)    IV

**16.** What is the molarity of a solution made by dissolving 8.56 g of sodium acetate in water and diluting to 750.0 mL?

| Formula | Molar Mass |
|---|---|
| $NaC_2H_3O_2$ | 82.03 g·mol$^{-1}$ |

   **(A)**   5.30 M      **(B)**   0.139 M      **(C)**   0.104 M      **(D)**   0.0783 M

**17.** A student wants to prepare 250. mL of 0.10 M NaCl solution. Which procedure is most appropriate?

| Formula | Molar Mass |
|---|---|
| NaCl | 58.4 g·mol$^{-1}$ |

   **(A)**   Add 5.84 g of NaCl to 250. mL of $H_2O$.

   **(B)**   Add 1.46 g of NaCl to 250. mL of $H_2O$.

   **(C)**   Dissolve 5.84 g of NaCl in 50 mL of $H_2O$ and dilute to 250. mL.

   **(D)**   Dissolve 1.46 g of NaCl in 50 mL of $H_2O$ and dilute to 250. mL

**18.** What volume of 12 M HCl solution is required to prepare exactly 500. mL of a 0.60 M HCl solution?

   **(A)**   10. mL      **(B)**   14 mL      **(C)**   25 mL      **(D)**   40. mL

**19.** In which set are the substances arranged in order of decreasing solubility in water?

   **(A)**   $Al(OH)_3 > Mg(OH)_2 > NaOH$      **(B)**   $BaSO_4 > CaSO_4 > MgSO_4$

   **(C)**   $CaCO_3 > NaHCO_3 > Na_2CO_3$      **(D)**   $AgCl > AgBr > AgI$

**20.** The solubility of a substance is 60 g per 100 mL of water at 15 °C. A solution of this substance is prepared by dissolving 75 g in 100 mL of water at 75 °C. The solution is then cooled slowly to 15 °C without any solid separating. The solution is

   **(A)**   supersaturated at 75 °C.      **(B)**   supersaturated at 15 °C.

   **(C)**   unsaturated at 15 °C.      **(D)**   saturated at 15 °C.

**21.** A mixture of 100 g of $K_2Cr_2O_7$ and 200 g of water is stirred at 60 °C until no more of the salt dissolves. The resulting solution is poured off, leaving the undissolved solid behind. The solution is now cooled to 20 °C. What mass of $K_2Cr_2O_7$ crystallizes from the solution during the cooling?

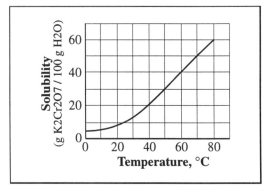

   **(A)**   24 g      **(B)**   31 g      **(C)**   43 g      **(D)**   62 g

**22.** What mass of water is needed to dissolve 292.5 g of NaCl to produce a 0.25 $m$ aqueous solution?

| Formula | Molar Mass |
|---|---|
| NaCl | 58.5 g·mol$^{-1}$ |

   **(A)**   20 kg      **(B)**   5.0 kg      **(C)**   0.80 kg      **(D)**   0.050 kg

**23.** What is the mole fraction of water in 200. g of 95% (by mass) ethanol, $C_2H_5OH$?

| Formula | Molar Mass |
|---|---|
| $C_2H_5OH$ | 46 g·mol$^{-1}$ |

   **(A)**   0.050      **(B)**   0.12      **(C)**   0.56      **(D)**   0.88

24. 800 g of ethanol, $C_2H_5OH$, was added to $8.0 \times 10^3$ g of water. How much would this lower the freezing point?

| **Freezing Point Depression Constant** |
| $K_f$ for water = 1.86 °C·$m^{-1}$ |

   (A)    3.2 °C       (B)    4.1 °C       (C)    8.2 °C       (D)    16 °C

25. The edge of a body-centered-cubic unit cell (which contains two atoms per unit cell) of an element **Y** was found to be $3.16 \times 10^{-8}$ cm. The density of the metal is 19.35 g·$cm^{-3}$. What is the approximate molar mass of **Y**?

   (A)    65.4 g·$mol^{-1}$    (B)    92.0 g·$mol^{-1}$    (C)    184 g·$mol^{-1}$    (D)    238 g·$mol^{-1}$

26. What is the number of nearest neighbors in a body-centered-cubic lattice?

   (A)    12       (B)    8       (C)    6       (D)    4

27. Carbon dioxide, $CO_2$, in the form of dry ice would be classified as

   (A)    an ionic solid.                    (B)    a polymeric solid.

   (C)    a molecular solid.                 (D)    a network solid.

28. Which diagrams represent pure substances?

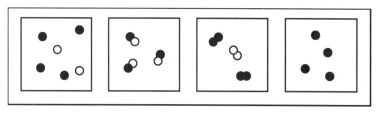

   (A)    **1** and **2**       (B)    **1** and **3**       (C)    **2** and **3**       (D)    **2** and **4**

29. Which diagram best represents the final system, if the pressure of the gas in this cylinder were doubled and the temperature increased from 200 K to 400 K?

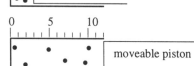

   (A)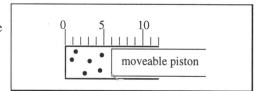

   (B)

   (C)

   (D)

30. An open-ended mercury manometer is used to measure the pressure exerted by a trapped gas as shown in the figure. Atmospheric pressure is 749 mmHg. What is the pressure of the trapped gas?

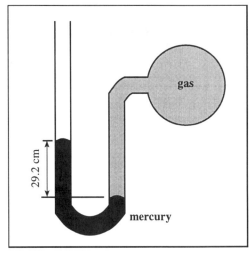

(A)     292 mmHg        (B)     457 mmHg

(C)     749 mmHg        (D)     1041 mmHg

## Answers to Study Questions

| | | | | | |
|---|---|---|---|---|---|
| 1. | B | 5. | D | 9. | A |
| 2. | B | 6. | A | 10. | C |
| 3. | A | 7. | A | | |
| 4. | C | 8. | D | | |

## Answers to Practice Questions

| | | | | | |
|---|---|---|---|---|---|
| 1. | B | 11. | A | 21. | D |
| 2. | D | 12. | C | 22. | A |
| 3. | B | 13. | C | 23. | B |
| 4. | B | 14. | B | 24. | B |
| 5. | C | 15. | C | 25. | C |
| 6. | B | 16. | B | 26. | B |
| 7. | B | 17. | D | 27. | C |
| 8. | A | 18. | C | 28. | D |
| 9. | A | 19. | D | 29. | C |
| 10. | D | 20. | B | 30. | D |

# *Energetics*

Energetics is a broad area that considers the nature of energy and its relationship with chemical and physical changes. Basic definitions of the vocabulary and units to express energy changes are important aspects of study, and must be mastered before building conceptual understanding. Heat flow under different conditions of pressure and temperature, calorimetry, use of tabulated enthalpy data for standard heats of formation, bond energies, and Hess's Law are typical topics that surround the First Law of Thermodynamics. An examination of spontaneous processes and a molecular interpretation of entropy lead to an understanding of the Second Law of Thermodynamics. Gibbs Free Energy and its relationship to equilibrium are the last areas included in the study of energetics.

| EN-1. | A 10.0 g sample of silver is heated to 100.0 °C and then added to 20.0 g of water at 23.0 °C in an insulated calorimeter. At thermal equilibrium, the temperature of the system was measured as 25.0 °C. What is the specific heat of silver? | Specific Heat Data | |
|---|---|---|---|
| | | $H_2O_{(l)}$ | 4.184 $J \cdot g^{-1} \cdot °C^{-1}$ |

    **(A)**    0.053 $J \cdot g^{-1} \cdot °C^{-1}$   **(B)**    0.22 $J \cdot g^{-1} \cdot °C^{-1}$   **(C)**    4.5 $J \cdot g^{-1} \cdot °C^{-1}$   **(D)**    8.4 $J \cdot g^{-1} \cdot °C^{-1}$

***Knowledge Required:*** (1) The principles of heat flow. (2) The meaning of the term specific heat. (3) The methods for carrying out calorimetry calculations.

***Thinking it Through:*** When the warmer silver is added to the cooler water in an insulated calorimeter, heat flow will take place from the silver to the water until thermal equilibrium is reached. Energy is conserved in this process and the heat lost by the silver is gained by the water, an illustration of the First Law of Thermodynamics. This law is also known as the Law of Conservation of Energy. The quantity of heat exchanged in this experiment is dependent on three factors—the amount of material, the heat exchanging capacity of the material as expressed by the specific heat, and the temperature change the material undergoes. This is the general relationship.

$$\text{heat} = \text{mass} \times \text{specific heat} \times \text{change in temperature}: \text{J} = \text{g} \times \frac{\text{J}}{\text{g} \cdot °C} \times (T_2 - T_1)\ °C$$

Given that the heat lost by the silver must be gained by the water, and assuming no heat loss, the relationship can be expressed in this manner. The symbol *s* represents the specific heat.

$$\text{heat lost} + \text{heat gained} = 0$$

$$10.0\,\text{g} \times s \times (100.0\ °C - 25.0\ °C) + 20.0\,\text{g} \times \frac{4.184\,\text{J}}{\text{g} \cdot °C} \times (23.0\ °C - 25.0\ °C) = 0$$

$$s = \frac{20.0\,\text{g} \times \dfrac{4.184\,\text{J}}{\text{g} \cdot °C} \times 2.0\ °C}{10.0\,\text{g} \times 75.0\ °C}$$

$$s = \frac{0.22\,\text{J}}{\text{g} \cdot °C}$$

This is choice **(B)**. Note that this is a reasonable value for silver compared with water's large specific heat. You know from experience that a silver spoon in a cup of hot coffee will quickly become too warm to touch. Water, with its large value for specific heat, warms and cools more slowly. If you solve the equation incorrectly in the last step and divide 750 by 167 rather than the correct 167 by 750, you will end up with choice **(C)**. If you have chosen choice **(A)**, you have neglected to use the specific heat of water. Choice **(D)** results from using an incorrect temperature change of 77 °C for both the silver and the water, neglecting the equilibrium temperature change.

| EN-2. | During the part of the cycle when heat is removed from the food compartment of an electric refrigerator, the refrigerant undergoes a change from a | | |
|---|---|---|---|
| | **(A)** liquid to a gas. | **(B)** | gas to a liquid. |
| | **(C)** liquid to a solid. | **(D)** | solid to a liquid. |

*Knowledge Required:* (1) The principles of heat flow. (2) The energy changes during phase changes.

*Thinking it Through:* If heat is *removed from* the food compartment, it must be *absorbed by* the refrigerant, so you will need to identify those changes in state that are *endothermic*. Both choices (**A**) and (**D**) describe changes in state that will absorb heat; choices (**B**) and (**C**) both involve condensation reactions that release heat so they can be eliminated from consideration. Choosing between (**A**) and (**D**) depends on the fact that it is far easier to circulate and compress fluids than it is to engineer an electric refrigerator based on the phase change from solid to liquid. This is, however, the phase change used in an old-fashioned icebox or your picnic cooler if you load it up with ice blocks or cubes. You may also know that generally the heat of vaporization for a substance is considerably higher than its heat of liquefaction, making the phase change from liquid to gas energetically favored.

---

**EN-3.** What is the net energy released in this reaction?

$$CH_4(g) + Cl_2(g) \rightarrow CH_3Cl(g) + HCl(g)$$

| Bond Energies, kJ·mol$^{-1}$ | | | |
|---|---|---|---|
| H–H | 431 | C–Cl | 327 |
| Cl–Cl | 239 | C–H | 410 |
| H–Cl | 427 | C–C | 335 |

| | (A) | 105 kJ | (B) | 188 kJ | (C) | 754 kJ | (D) | 1403 kJ |
|---|---|---|---|---|---|---|---|---|

*Knowledge Required:* (1) The meaning of bond energy. (2) The energy changes that take place when bonds are broken and formed.

*Thinking it Through:* Bond energy, also called by the more descriptive name of bond-dissociation energy, is the energy required to break a bond when the substance is in the gas phase. It *always requires* energy to break bonds, so values for bond energies are always tabulated with a positive sign implied. However, when a given bond forms in the gas phase, it will *always release* that same amount of bond energy. It is the interplay between the bonds broken and the bonds formed that will determine if the reaction overall is endothermic or, as is the case here, exothermic.

Break one C–H bond    Break one Cl–Cl bond    Form one C–Cl bond    Form one H–Cl bond

There are two bonds to be broken. For the C–H bond, 410 kJ of energy are required per mole. For the Cl–Cl bond, 239 kJ are required per mole. The total energy that must be used is 649 kJ.

There are two bonds that are formed. For the C–Cl bond, 327 kJ of energy are released. For the H–Cl bond, 427 kJ are released. The total energy released is 754 kJ.

Overall, more energy is released in forming bonds than was absorbed in breaking bonds. This defines an exothermic reaction and there is a net energy release of (754 kJ – 649 kJ), which equals 105 kJ. This is choice (**A**). Choosing (**C**) means that only the energy released was considered, not the *net* energy released. Choice (**B**) has only focused on the energy required for breaking the Cl–Cl bond relative to that released in forming the H–Cl bond. The other bond energy changes have been ignored. Choice (**D**) is the inappropriate combination of both the energy required and the energy released.

*Notes:* Bond energies are average experimental values and there is some variation in their values in different questions. Asking for the enthalpy change or $\Delta H_{rxn}$ is equivalent to asking for the energy change for this gas phase reaction. Remember, however, that values of $\Delta H_{rxn}$ must follow sign conventions. A positive sign is associated with an endothermic reaction and a negative sign with an exothermic reaction.

---

**EN-4.** What mass of benzene, $C_6H_6(l)$, must be burned in a bomb calorimeter to raise its temperature by 15 °C?

| Thermodynamic Data | |
|---|---|
| $\Delta H^0_{combustion}(C_6H_6(l))$ | –41.9 kJ·g$^{-1}$ |
| Calorimeter Constant | 1.259 kJ·°C$^{-1}$ |

| | (A) | 0.45 g | (B) | 2.8 g | (C) | 3.5 g | (D) | 35 g |
|---|---|---|---|---|---|---|---|---|

*Knowledge Required:* (1) The principles of bomb calorimetry. (2) The meaning of heats of combustion.

*Thinking it Through:* A bomb calorimeter is an insulated container that contains a reaction chamber designed to withstand high pressures. The reaction chamber, or bomb, is surrounded by water that absorbs the heat of the reaction. This experimental device is often used to study reactions at constant volume, particularly the combustion reactions of organic compounds. A sample of the compound being burned is placed in the reaction chamber that is then sealed and pressurized with oxygen. An electric current is passed through a thin wire that gets hot enough to initiate the combustion reaction. Heat is given off during the reaction and that heat is absorbed by the surroundings, causing the measured temperature of the water to rise. In this case, the heat capacity of the calorimeter is given, having been previously determined experimentally. This is called the calorimeter constant. It means that it takes 1.259 kJ of energy to raise the temperature of the reaction chamber's surroundings by one degree Celsius. The value of the heat of combustion is also given, showing that benzene will give off 4.19 kJ of heat energy for every gram of benzene burned. The heat lost by the benzene burning is equal to the heat gained by the surroundings, assuming no heat loss. Let $x$ be the number of grams of benzene.

$$\text{heat evolved} + \text{heat gained} = 0$$

$$x \times \frac{41.9\,\text{kJ}}{g} + 15\,^{\circ}\text{C} \times \frac{1.259\,\text{kJ}}{^{\circ}\text{C}} = 0$$

$$x = 0.45\,\text{g}$$

This is choice (**A**). If you neglected to use the calorimeter constant, you might choose choice (**B**) in error. If you solved this equation incorrectly, you might choose choice (**C**). Choice (**D**) could be chosen incorrectly if the 15° temperature change was ignored.

*Notes:* Often problems of this type will involve the heat of combustion in units of kJ per mole rather than per gram. The molar mass of the substance undergoing combustion can be used to make the necessary conversions.

---

**EN-5.** Use the heats of formation, $\Delta H_f^0$, in the table to determine $\Delta H$ for this reaction.

$$2C_2H_2(g) + 5O_2(g) \rightarrow 4CO_2(g) + 2H_2O(g)$$

| Compound | $\Delta H_f^0$, kJ·mol$^{-1}$ |
|---|---|
| $C_2H_2(g)$ | +227 |
| $H_2O(g)$ | –242 |
| $CO_2(g)$ | –393 |

(A)  –1602 kJ    (B)  –1829 kJ    (C)  –2283 kJ    (D)  –2510 kJ

*Knowledge Required:* (1) The meaning of standard heats of formation. (2) The methods of using heats of formation to calculate heats of reaction.

*Thinking it Through:* The enthalpy change for any reaction occurring at constant pressure can be calculated from the given or tabulated values of standard heats of formation for both reactants and products. It is necessary to sum the heats of formation for all reaction products, being sure to multiply each molar heat of formation by the coefficient of that substance given in the balanced chemical equation. From this total, it is necessary to subtract the sum of the heats of formation for all of the reactants, again multiplying by the coefficients in the balanced chemical equation. This is the generalization for the process.

$$\Delta H_{rxn}^0 = \sum n\,\Delta H_f^0(\text{products}) - \sum m\,\Delta H_f^0(\text{reactants})$$

The symbol $\sum$ (sigma), means "the sum of" and $n$ and $m$ are the total of the coefficients of the products and reactants, respectively. To apply this generalization in this question, use the given values from the table and recall that the standard enthalpy of formation for any element in its standard state is zero. Oxygen gas is an element in its standard state, so no value is given in the table.

$$\Delta H_{rxn}^0 = [(4\,\Delta H_f^0(CO_2) + 2\,\Delta H_f^0(H_2O)] - [2\,\Delta H_f^0(C_2H_2) + 5\,\Delta H_f^0(O_2)]$$

$$\Delta H_{rxn}^0 = [(4 \times (-393\,\text{kJ}) + 2 \times (-242\,\text{kJ})] - [2 \times (+227\,\text{kJ}) + 5 \times (0\,\text{kJ})]$$

$$\Delta H_{rxn}^0 = -2510\,\text{kJ}$$

This is choice (**D**). Common errors in this type of question are forgetting to use the coefficients or reversing the signs and subtracting the products from the reactants. Various coefficient and sign errors lead to the incorrect answers in choices (**A**) through (**C**).

**EN-6**     Given:

$$\text{Equation 1: } SO_{2(g)} \rightarrow S_{(s)} + O_{2(g)} \qquad \Delta H = +300 \text{ kJ}$$

$$\text{Equation 2: } 2SO_{2(g)} + O_{2(g)} \rightarrow 2SO_{3(g)} \qquad \Delta H = -200 \text{ kJ}$$

Use this information to calculate the heat of formation of $SO_{3(g)}$.

(A)    $-500$ kJ·mol$^{-1}$   (B)    $-400$ kJ·mol$^{-1}$   (C)    $+100$ kJ·mol$^{-1}$   (D)    $+200$ kJ·mol$^{-1}$

---

***Knowledge Required:*** (1) The definition of enthalpy change, $\Delta H$. (2) The definition of heat of formation, $\Delta H_f^0$
(3) The meaning of Hess's Law.

***Thinking it Through:*** The change in enthalpy for a chemical reaction is equal to the heat that is absorbed or released in a chemical reaction taking place at constant temperature and pressure. Enthalpy is an example of a *state function*, one that depends only on the initial and final states of the reactants and products. Enthalpy changes also depend on the amount of material present. Equations for two chemical reactions and their accompanying enthalpy changes are given in this question. However, what is asked for is the *heat of formation*, $\Delta H_f^0$, for $SO_{3(g)}$.

$$S_{(s)} + \frac{3}{2}O_{2(g)} \rightarrow SO_{3(g)} \qquad \Delta H_f^0 = ? \text{ kJ·mol}^{-1}$$

This will be the enthalpy change associated with the formation of $SO_3$ from its elements, all in their standard states at 25 °C and one atmosphere of pressure. Often it is difficult to carry out such a seemingly straightforward reaction and measure its heat. It is often more convenient to algebraically combine equations and their enthalpies. Chemical equations and the enthalpy changes that accompany them can be combined algebraically only because $\Delta H$ does not depend on the path from a set of reactants to a set of products. Rather, it only depends on the change from initial to final states. *Hess's Law* states that if a reaction is carried out in a series of steps, $\Delta H$ for the overall reaction can be calculated from the sum of the enthalpy changes for the individual steps. Given experimental data can be changed in any way that is algebraically convenient, so long as the accompanying enthalpy change is also adjusted. For example, if an equation is reversed, the sign of the enthalpy change must be reversed. If an equation is multiplied by a factor, the enthalpy must be multiplied by that same factor. Applying this in this question, and keeping an eye on the desired equation, the first reaction will need to be reversed so the sign of the enthalpy change is also reversed. The second reaction and its enthalpy change is multiplied by 1/2.

**Equation 1:** $SO_{2(g)} \rightarrow S_{(s)} + O_{2(g)}$     $\Delta H = +300$ kJ     becomes $S_{(s)} + O_{2(g)} \rightarrow SO_{2(g)}$     $\Delta H = -300$ kJ

**Equation 2:** $2SO_{2(g)} + O_{2(g)} \rightarrow 2SO_{3(g)}$     $\Delta H = -200$ kJ     becomes $SO_{2(g)} + \frac{1}{2}O_{2(g)} \rightarrow SO_{3(g)}$   $\Delta H = -100$ kJ

Adding the transformed equations together gives:

$$S_{(s)} + O_{2(g)} + SO_{2(g)} + \frac{1}{2}O_{2(g)} \rightarrow SO_{2(g)} + SO_{3(g)} \qquad \Delta H = (-300 \text{ kJ}) + (-100 \text{ kJ}) = -400 \text{ kJ}$$

Combining like terms and canceling substances that appear on both sides of the equation gives the target equation.

$$S_{(s)} + \frac{3}{2}O_{2(g)} \rightarrow SO_{3(g)} \qquad \Delta H_f = -400 \text{ kJ·mol}^{-1}$$

This is choice (**B**). You might pick choice (**A**) in error if you forget to reverse the sign of the enthalpy for Equation 1. If you forgot the sign reversal but also neglected to cut the enthalpy associated with the second equation in half, then you might have chosen choice (**C**). Choice (**D**) might be selected if the second equation were correctly cut in half, but also reversed and added to **Equation 1**.

| | | | |
|---|---|---|---|
| **EN-7.** | Which change is likely to be accompanied by an increase in entropy? | | |

    **(A)**    $N_2(g) + 3H_2(g) \rightarrow 2NH_3(g)$ at 25 °C    **(B)**    $Ag^+(aq) + Cl^-(aq) \rightarrow AgCl(s)$ at 25 °C

    **(C)**    $CO_2(s) \rightarrow CO_2(g)$ at –70 °C    **(D)**    $H_2O(g) \rightarrow H_2O(l)$ at 100 °C

***Knowledge Required:*** (1) The meaning of the term *entropy*. (2) The interpretation of chemical equations.

***Thinking it Through:*** Entropy is a thermodynamic quantity that expresses the disorder or randomness in a system. Considering whether the reactants or the products are more randomly arranged can identify the change accompanied by the greatest increase in entropy. In choice **(A)**, all of the particles are present in gas phase and there is a change from four moles of gas of diatomic gases to only two moles of more complex ammonia molecules. This system is becoming more ordered, a process likely to produce a decrease in entropy. Choice **(B)** is forming a highly ordered crystal from ions in solution, so once more the process is predicted to be accompanied by a decrease in entropy. Choice **(C)**, the correct choice, will have an increase in entropy, for solid carbon dioxide is changing to gaseous carbon dioxide, a highly disorganized state. $CO_2$ molecules in the gas phase have many more opportunities to vibrate and will have much higher entropy than the condensed solid phase. Choice **(D)** represents a condensation from the gas to the liquid states; this again will be a decrease, not an increase in randomness. Do not be misled by the temperatures that are given. Although the term $T\Delta S$ gives the total energy contribution in a chemical change from the temperature multiplied by the change in entropy, it is only the change in entropy that is asked in this question.

| | |
|---|---|
| **EN-8.** | Under which circumstance would the free energy change for a reaction be relatively temperature independent? |

    **(A)**    $\Delta H$ is negative.    **(B)**    $\Delta H$ is positive.

    **(C)**    $\Delta S$ has a large positive value.    **(D)**    $\Delta S$ has a small value of either sign.

***Knowledge Required:*** (1) The meaning of the term *free energy*. (2) The relationship of enthalpy and entropy to free energy.

***Thinking it Through:*** The thermodynamic function known as the Gibbs free energy is the relationship that predicts whether a chemical reaction is spontaneous. It depends on two values, the enthalpy of reaction and the product of the temperature and the entropy change for the reaction. For a reaction taking place at constant temperature, the relationship becomes $\Delta G = \Delta H - T\Delta S$. In general, the values of $\Delta H$ are in kJ and the values of $\Delta S$ are in joules per Kelvin, $J \cdot K^{-1}$. This means that unless the temperature is very high, so that the product of $T\Delta S$ becomes larger than the magnitude of $\Delta H$, the spontaneity of a reaction is controlled by the sign of $\Delta H$. If the sign of $\Delta G$ is negative, the reaction will be spontaneous in the forward direction. If $\Delta G$ is equal to zero, the reaction is at equilibrium because there is no driving force in either direction. If the sign of $\Delta G$ is positive, the reaction as written will be nonspontaneous. This question asks for the relationship of the size of $\Delta H$ to that of $T\Delta S$ in order to make a qualitative judgement about the value of the free energy. In choice **(A)**, even if $\Delta H$ is negative, the value of $\Delta G$ might become positive if the temperature were high and/or the entropy change were large enough. In choice **(B)**, $\Delta H$ is positive, which usually predicts a nonspontaneous reaction unless there is significant entropy change and high temperature conditions. In choice **(C)**, there is still the possibility that the value of $T\Delta S$ could control the reaction, even if the value of $\Delta H$ were negative. Only choice **(D)** gives the possibility of a relatively temperature independent value for the free energy change. If $\Delta S$ has a small value, no matter what the sign, then the product of the temperature and the entropy change will be a small number compared to the size of the enthalpy change.

| EN-9 | Which of the relationships is(are) true about water boiling in a container that is open to the atmosphere? | | Relationships | |
|---|---|---|---|---|
| | | | I | $\Delta H = 0$ |
| | | | II | $\Delta S = 0$ |
| | | | III | $\Delta G = 0$ |

(A)  I only       (B)  III only

(C)  I and II only       (D)  II and III only

*Knowledge Required:* (1) The meaning of the thermodynamic functions of enthalpy, entropy, and free energy. (2) The changes in thermodynamic relationships during change in state.

*Thinking it Through:* The normal boiling point of a liquid is the temperature at which the vapor pressure of the liquid is equal to 760 mmHg or 1 atmosphere pressure. When water is boiling, there is equilibrium between the rate at which water molecules are escaping the surface and the rate at which molecules are condensing. Considerable energy must be supplied to water for its molecules to have enough energy to escape the surface of the liquid, so $\Delta H$ cannot be equal to zero. This eliminates both choices (A) and (C) from consideration. The molecules in the vapor state are more random than the more ordered molecules in the liquid state, so $\Delta S$ also cannot be equal to zero. This eliminates choice (C) again, and now also choice (D). Choice (B) is correct because it identifies the change in free energy as zero, a characteristic of an equilibrium state.

---

**EN-10**  For this reaction at 25 °C, $\Delta H° = -1854$ kJ and $\Delta S° = -236$ J·K$^{-1}$.

$$CH_3COCH_{3(g)} + 4O_{2(g)} \rightarrow 3CO_{2(g)} + 3H_2O_{(l)}$$

What is the value of $\Delta G°$ for this reaction?

(A)  −1784 kJ     (B)  −1848 kJ     (C)  −1924 kJ     (D)  68500 kJ

*Knowledge Required:* (1) The calculation of free energy from enthalpy and entropy data. (2) The correct use of units.

*Thinking it Through:* The free energy change for any reaction can be calculated using $\Delta G = \Delta H - T\Delta S$. To do this successfully, take careful note of the units.

$$\Delta G = -1854 \text{ kJ} - 298 \text{ K} \times \frac{-263 \text{ J}}{\text{K}} \times \frac{1 \text{ kJ}}{1000 \text{ J}}$$

$$\Delta G = -1854 \text{ kJ} + 70 \text{ kJ}$$

$$\Delta G = -1784 \text{ kJ}$$

This is choice (A). Common errors in this type of calculation include forgetting to change the temperature from 25 °C to 298 K, resulting in choice (B). It is also easy to make a sign error in the second part of the expression, carelessly making the product of two negatives values still a negative value; this produces the incorrect choice (C). If you neglect to change joules to kilojoules, then you will combine 70,000 J with kJ and obtain the large value in choice (D). This careless error should be easy to spot, for remember that for most combustion reactions, the magnitude of the enthalpy term provides the driving force for the reaction.

## Practice Questions

1. When a 45.0 g sample of an alloy at 100.0 °C is dropped into 100.0 g of water at 25.0 °C, the final temperature is 37.0 °C. What is the specific heat of the alloy?

| Specific Heat Data | |
|---|---|
| $H_2O(l)$ | 4.184 J·g$^{-1}$·°C$^{-1}$ |

   (A)     0.423 J·g$^{-1}$·°C$^{-1}$   (B)     1.77 J·g$^{-1}$·°C$^{-1}$   (C)     9.88 J·g$^{-1}$·°C$^{-1}$   (D)     48.8 J·g$^{-1}$·°C$^{-1}$

2. When 68.00 J of energy are added to a sample of gallium that is initially at 25.0 °C, the temperature rises to 38.0 °C. What is the volume of the sample?

| Data for Gallium, Ga | |
|---|---|
| specific heat | 0.372 J·g$^{-1}$·°C$^{-1}$ |
| density | 5.904 g·cm$^{-3}$ |

   (A)     2.38 cm$^3$     (B)     4.28 cm$^3$     (C)     14.1 cm$^3$     (D)     31.0 cm$^3$

3. A student mixes 100 mL of 0.50 M NaOH with 100 mL of 0.50 M HCl in a Styrofoam® cup and observes a temperature increase of $\Delta T_1$. When she repeats this experiment using 200 mL of each solution, she observes a temperature change of $\Delta T_2$. If no heat is lost to the surroundings or absorbed by the Styrofoam cup, what is the relationship between $\Delta T_1$ and $\Delta T_2$?

   (A)     $\Delta T_2 = 4\,\Delta T_1$     (B)     $\Delta T_2 = 2\,\Delta T_1$     (C)     $\Delta T_2 = 0.5\,\Delta T_1$     (D)     $\Delta T_2 = \Delta T_1$

4. When a material in the liquid state is vaporized and then condensed to a liquid, the steps in the process are, respectively,

   (A)     exothermic and exothermic.          (B)     exothermic and endothermic.

   (C)     endothermic and exothermic.          (D)     endothermic and endothermic.

5. When $Na_2S_2O_3\cdot3H_2O$ dissolves in water, the solution gets cold. Which diagram best represents the change in enthalpy for the contents of the flask?

   (A)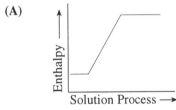

   (B)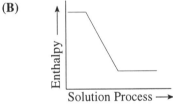

   (C)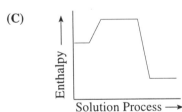

   (D)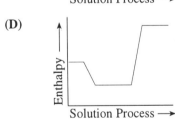

6. Calculate $\Delta H^{\circ}$ = for the chemical reaction

$$Cl_2(g) + F_2(g) \rightarrow 2ClF(g)$$

| Bond Energies, kJ·mol$^{-1}$ | |
|---|---|
| F–F | 159 |
| Cl–Cl | 243 |
| Cl–F | 255 |

   (A)     –147 kJ     (B)     –108 kJ     (C)     +171 kJ     (D)     +912 kJ

7. More heat is derived from cooling one gram of steam at 100 °C to water at 50 °C than from cooling one gram of liquid water at 100 °C to 50 °C because

    (A)    the steam is hotter than the water.

    (B)    the steam occupies a greater volume than the water.

    (C)    the density of water is greater than that of steam.

    (D)    the heat of condensation is evolved.

8. Calculate $\Delta H$ for this gas phase reaction.

    $$NH_3(g)+ Cl_2(g) \rightarrow NH_2Cl(g)+ HCl(g)$$

| Bond Energies, kJ·mol⁻¹ | | | |
|---|---|---|---|
| N–H | 389 | H–Cl | 427 |
| N–Cl | 201 | Cl–Cl | 243 |

    (A)    –632 kJ     (B)    –188 kJ     (C)    +4 kJ     (D)    +431 kJ

9. Use the bond energies in the table to determine $\Delta H$ for the formation of hydrazine, $N_2H_4$, from nitrogen and hydrogen according to this equation.

    $$N_2(g) + 2H_2(g) \rightarrow N_2H_4(g)$$

| Bond Energies, kJ·mol⁻¹ | | | |
|---|---|---|---|
| N–N | 159 | H–H | 436 |
| N=N | 201 | H–N | 389 |
| N≡N | 941 | | |

    (A)    –711 kJ     (B)    –98 kJ     (C)    +98 kJ     (D)    +711 kJ

10. A 1.00 g sample of $NH_4NO_3$ is decomposed in a bomb calorimeter. The temperature increases by 6.12 °C. What is the molar heat of decomposition for ammonium nitrate?

| Table of Data | |
|---|---|
| Molar mass, $NH_4NO_3$ | 80.0 g·mol⁻¹ |
| Calorimeter Constant | 1.23 kJ·°C⁻¹ |

    (A)    –602 kJ·mol⁻¹    (B)    –398 kJ·mol⁻¹    (C)    7.53 kJ·mol⁻¹    (D)    164 kJ·mol⁻¹

11. In a bomb calorimeter, reactions are carried out at

    (A)    constant pressure.          (B)    constant volume.

    (C)    1 atm pressure and 25 °C.    (D)    1 atm pressure and 0 °C.

12. The standard enthalpy of formation, $\Delta H_f^\circ$, for nitrogen(IV) oxide, $NO_2$, is the enthalpy change for which reaction?

    (A)    $N(g) + 2O(g) \rightarrow NO_2(g)$

    (B)    $\frac{1}{2}N_2(g) + O_2(g) \rightarrow NO_2(g)$

    (C)    $\frac{1}{2}N_2O_4(g) \rightarrow NO_2(g)$

    (D)    $NO(g) + \frac{1}{2}O_2(g) \rightarrow NO_2(g)$

13. The combustion of ammonia is represented by this equation.

    $$4NH_3(g) + 5O_2(g) \rightarrow 4NO(g) + 6H_2O(g) \quad \Delta H_{rxn}^\circ = -904.8 \text{ kJ}$$

    What is the enthalpy of formation of $NH_3(g)$?

| Enthalpy of Formation Data | |
|---|---|
| $NO(g)$ | +90.4 kJ·mol⁻¹ |
| $H_2O(g)$ | –241.8 kJ·mol⁻¹ |

    (A)    –449 kJ·mol⁻¹    (B)    –46.1 kJ·mol⁻¹    (C)    –184 kJ·mol⁻¹    (D)    227 kJ·mol⁻¹

14. The standard enthalpy change for the formation of silver chloride from its elements is _____than the enthalpy change for the formation of silver bromide.

| Standard Enthalpies of Formation, 298 K | |
|---|---|
| AgCl(s) | $-127.1$ kJ·mol$^{-1}$ |
| AgBr(s) | $-100.4$ kJ·mol$^{-1}$ |

(A)  more exothermic

(B)  less exothermic

(C)  more endothermic

(D)  less endothermic

15. What is the value of $\Delta H^\circ$ for this reaction?

$$C_4H_4(g) + 2H_2(g) \rightarrow C_4H_8(g)$$

| Enthalpies of Combustion | |
|---|---|
| $C_4H_4(g)$ | $-2341$ kJ·mol$^{-1}$ |
| $H_2(g)$ | $-286$ kJ·mol$^{-1}$ |
| $C_4H_8(g)$ | $-2755$ kJ·mol$^{-1}$ |

(A)  $-158$ kJ   (B)  $-128$ kJ   (C)  $+128$ kJ   (D)  $+158$ kJ

16. Given these values of $\Delta H^\circ$:

$$CS_2(l) + 3O_2(g) \rightarrow CO_2(g) + 2SO_2(g) \qquad \Delta H^\circ = -1077 \text{ kJ}$$

$$H_2(g) + O_2(g) \rightarrow H_2O_2(l) \qquad \Delta H^\circ = -188 \text{ kJ}$$

$$H_2(g) + \frac{1}{2}O_2(g) \rightarrow H_2O(l) \qquad \Delta H^\circ = -286 \text{kJ}$$

What is the value of $\Delta H^\circ$ for this reaction?

$$CS_2(l) + 6H_2O_2(l) \rightarrow CO_2(g) + 6 H_2O(l) + 2SO_2(g$$

(A)  $-1175$ kJ·mol$^{-1}$   (B)  $-1151$ kJ·mol$^{-1}$   (C)  $-1665$ kJ·mol$^{-1}$   (D)  $-3921$ kJ·mol$^{-1}$

17. Use the standard enthalpies of formation in the table to calculate $\Delta H^\circ$ for this reaction.

$$2CrO_4{}^{2-}(aq) + 2H^+(aq) \rightarrow Cr_2O_7{}^{2-}(aq) + H_2O(l)$$

| Substance | $\Delta H_f^\circ$, kJ·mol$^{-1}$ |
|---|---|
| $CrO_4{}^{2-}(aq)$ | $-881.2$ |
| $Cr_2O_7{}^{2-}(aq)$ | $-1490.3$ |
| $H^+(aq)$ | $0$ |
| $H_2O$ (l) | $-285.8$ |

(A)  $+272.1$ kJ   (B)  $+13.7$ kJ   (C)  $-13.7$ kJ   (D)  $-272.1$ kJ

18. What is the value of $\Delta H^\circ$ for this reaction?

$$3H_2(g) + O_3(g) \rightarrow 3H_2O(l)$$

| Reaction | $\Delta H$, kJ |
|---|---|
| $H_2(g) + 1/2O_2(g) \rightarrow H_2O(l)$ | $-286$ kJ |
| $3O_2(g) \rightarrow 2O_3(g)$ | $+271$ kJ |

(A)  $-15$ kJ   (B)  $-558$ kJ   (C)  $-722$kJ   (D)  $-994$ kJ

19. Calculate the enthalpy of combustion of ethylene, $C_2H_4$, at 25 °C and one atmosphere pressure.

$$C_2H_4(g) + 3O_2(g) \rightarrow 2CO_2(g) + 2H_2O(l)$$

| Compound | $\Delta H_f^\circ$, kJ·mol$^{-1}$ |
|---|---|
| $C_2H_4(g)$ | $+52.3$ |
| $H_2O(l)$ | $-285.8$ |
| $CO_2(g)$ | $-393.5$ |

(A)  $-1411$ kJ   (B)  $-1254$ kJ   (C)  $-732$kJ   (D)  $-627$ kJ

20. Use the given heats of formation to calculate the enthalpy change for this reaction.

    $$B_2O_3(s) + 3COCl_2(g) \rightarrow 2BCl_3(g) + 3CO_2(g)$$

| Compound | $\Delta H_f^o$, kJ·mol$^{-1}$ |
|---|---|
| $B_2O_3(s)$ | −1272.8 |
| $COCl_2(g)$ | −218.8 |
| $BCl_3(g)$ | −403.8 |
| $CO_2(g)$ | −393.5 |

    (A)  694.3 kJ    (B)  354.9 kJ    (C)  −58.9 kJ    (D)  −3917.3 kJ

21. Using the given thermochemical data, what is $\Delta H^o$ for this reaction?

    $$2CH_3OH(l) + O_2(g) \rightarrow HC_2H_3O_2(l) + 2H_2O(l)$$

| Compound | $\Delta H_f^o$, kJ·mol$^{-1}$ |
|---|---|
| $CH_3OH(l)$ | −238 |
| $HC_2H_3O_2(l)$ | −487 |
| $H_2O(l)$ | −286 |

    (A)  583 kJ    (B)  535 kJ    (C)  −583 kJ    (D)  −535 kJ

22. What is the enthalpy change for this reaction?

    $$Hg(l) + 2Ag^+(aq) \rightarrow Hg^{2+}(aq) + 2Ag(s)$$

| Compound | $\Delta H_f^o$, kJ·mol$^{-1}$ |
|---|---|
| $Ag^+(aq)$ | +105.6 |
| $Hg^{2+}(aq)$ | +171.1 |

    (A)  +65.5 kJ    (B)  +40.1 kJ    (C)  −40.1 kJ    (D)  −65.5 kJ

23. For which of these processes is the value of $\Delta S$ expected to be negative?
    I.   Sugar is dissolved in water.
    II.  Steam is condensed.
    III. $CaCO_3$ is decomposed into CaO and $CO_2$.

    (A)  **I** only    (B)  **I** and **III** only
    (C)  **II** only    (D)  **II** and **III** only

24. For which process is the entropy change per mole the largest at constant temperature?

    (A)  $H_2O(l) \rightarrow H_2O(g)$    (B)  $H_2O(s) \rightarrow H_2O(g)$
    (C)  $H_2O(s) \rightarrow H_2O(l)$    (D)  $H_2O(l) \rightarrow H_2O(s)$

25. In which process is entropy *decreased*?

    (A)  dissolving sugar in water    (B)  expanding a gas
    (C)  evaporating a liquid    (D)  freezing water

26. When a liquid evaporates, which is true about the signs of the enthalpy and entropy changes?

    (A)  $\Delta H$ is positive, $\Delta S$ is positive    (B)  $\Delta H$ is positive, $\Delta S$ is negative
    (C)  $\Delta H$ is negative, $\Delta S$ is positive    (D)  $\Delta H$ is negative, $\Delta S$ is negative

27. A particular chemical reaction has a negative $\Delta H$ and a negative $\Delta S$. Which statement is correct?

    (A)  The reaction is spontaneous at all temperatures.
    (B)  The reaction is nonspontaneous at all temperatures.
    (C)  The reaction becomes spontaneous as temperature increases.
    (D)  The reaction becomes spontaneous as temperature decreases.

28. When solid $NH_4NO_3$ is dissolved in water at 25 °C, the temperature of the solution decreases. What is true about the signs of $\Delta H$ and $\Delta S$ for this process?

    (A)    $\Delta H$ is negative, $\Delta S$ is positive       (B)    $\Delta H$ is negative, $\Delta S$ is negative

    (C)    $\Delta H$ is positive, $\Delta S$ is positive       (D)    $\Delta H$ is positive, $\Delta S$ is negative

29. For the process $O_2(g) \rightarrow 2O(g)$, $\Delta H° = +498$ kJ. What would be predicted for the sign of $\Delta S_{rxn}$ and the conditions under which this reaction would be spontaneous?

    | | $\Delta S_{rxn}$ | Spontaneous |
    |---|---|---|
    | (A) | positive | at low temperatures only |
    | (B) | positive | at high temperatures only |
    | (C) | negative | at high temperatures only |
    | (D) | negative | at low temperatures only |

30. For which reaction, carried out at standard conditions, would both the enthalpy and entropy changes drive the reaction in the same direction?

    (A)    $2H_2(g) + O_2(g) \rightarrow 2H_2O(l)$        $\Delta H = -571.1$ kJ

    (B)    $2Na(s) + Cl_2(g) \rightarrow 2NaCl(s)$        $\Delta H = -822.0$ kJ

    (C)    $N_2(g) + 2O_2(g) \rightarrow 2NO_2(g)$        $\Delta H = +67.7$ kJ

    (D)    $2NH_3(g) \rightarrow N_2(g) + 3H_2(g)$ (l)        $\Delta H = +92.4$ kJ

## Answers to Study Questions

| | | | | | |
|---|---|---|---|---|---|
| 1. | B | 5. | D | 9. | B |
| 2. | A | 6. | B | 10. | A |
| 3. | A | 7. | C | | |
| 4. | A | 8. | D | | |

## Answers to Practice Questions

| | | | | | |
|---|---|---|---|---|---|
| 1. | B | 11. | B | 21. | C |
| 2. | A | 12. | B | 22. | C |
| 3. | D | 13. | B | 23. | C |
| 4. | C | 14. | A | 24. | B |
| 5. | A | 15. | A | 25. | D |
| 6. | B | 16. | C | 26. | A |
| 7. | D | 17. | C | 27. | D |
| 8. | C | 18. | D | 28. | C |
| 9. | C | 19. | A | 29. | B |
| 10. | A | 20. | C | 30. | C |

# Dynamics

A puzzling aspect of chemical reactions to many beginning students is the fact that many reactions that have a large thermodynamic driving force take place slowly or not at all. On the other hand, many other reactions that have very little thermodynamic driving force take place rapidly. The change in *Gibbs energy* (also called *free energy*) of a chemical reaction is the measure of the thermodynamic driving force. The Gibbs energy only tells us about the relationship of the initial and final energy states. It tells us nothing about the energy terrain in moving from reactants to products. The study of *reaction dynamics* (also called *reaction kinetics* or *rates of reaction*) reveals something of the energy topography in moving from reactants to products.

Once the rate of a chemical reaction has been thoroughly studied, reasonable conclusions can be drawn concerning the sequence of events (collisions, attractive interactions, repulsive interactions, bond breaking, bond making) that lead to the observed products. Developing the chronology of intermediate events, referred to as *reaction mechanisms*, are the ultimate goals of studies of reaction dynamics.

The basic experimental techniques for studying reaction dynamics involve either (1) instantaneous rate determinations at various known concentrations of reactants, usually when the reaction is just begun, or (2) determining the relationship between concentration and time.

## Study Questions

**DY-1.** The rate law for the reaction

$$A + B \rightarrow C + D$$

is first order in [A] and second order in [B]. If [A] is halved and [B] is doubled, the rate of the reaction will

| | | | |
|---|---|---|---|
| **(A)** | remain the same. | **(B)** | be increased by a factor of 2. |
| **(C)** | be increased by a factor of 4. | **(D)** | be increased by a factor of 8. |

*Knowledge Required:* (1) The meanings of the terms rate law, first order, and second order. (2) The algebraic form of the rate equation.

*Thinking it Through:* For this reaction, the rate equation is

$$\text{rate} = k\,[A][B]^2$$

Here, we are comparing an initial rate ($\text{rate}_1$) with a later rate ($\text{rate}_2$), and the responses are expressed as ratios of the two rates. The problem also specifies that $[A]_2 = 1/2[A]_1$ and $[B]_2 = 2[B]_1$.

$$\frac{\text{rate}_2}{\text{rate}_1} = \frac{k[A]_2[B]_2^2}{k[A]_1[B]_1^2} = \frac{k(\tfrac{1}{2}[A]_1) \times (2[B]_1)^2}{k[A]_1[B]_1^2}$$

$$\frac{\text{rate}_2}{\text{rate}_1} = \frac{(\tfrac{1}{2}) \times (2)^2}{1} = 2$$

The correct answer is choice (**B**). Choice (**A**) results if the square term is applied only to $[B]_1$ (rather than $2[B]_1$) in the numerator. Choice (**C**) results when the 1/2 term is missed. Choice (**D**) results when the 1/2 term is placed in the denominator rather than the numerator.

**DY-2.** Which line in the diagram represents the activation energy for a forward reaction?

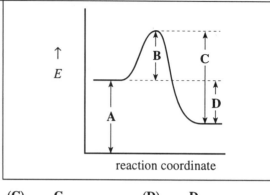

(A)   A          (B)   B          (C)   C          (D)   D

*Knowledge Required:* (1) Interpretation of energy *vs.* reaction coordinate plots. (2) Meaning of activation energy for forward and reverse reactions.

*Thinking it Through:* Reaction energy diagrams are drawn to represent energy change for the forward reaction as moving from left to right along the reaction coordinate. The energy change for the reverse reaction is interpreted as moving from right to left.

Energy changes are always measured relative to the *initial energy of reactants*, which is the beginning point of the energy profile. The activation energy is the height of the energy barrier relative to that base line, which is choice **(B)** in the diagram. Choice **(A)** is a meaningless distance, since the *y* intersection with the *x* axis is not defined. Choice **(C)** is the activation energy for the *reverse* reaction. Choice **(D)** is the net energy change for the reaction.

---

**DY-3.** Initial rate data for the reaction

$$2N_2O_5(g) \rightleftharpoons 4NO_2(g) + O_2(g)$$

are given in the table. What is the rate law for this reaction?

| Experiment | $[N_2O_5]$ | $[O_2]$ | Rate in M·s$^{-1}$ |
|---|---|---|---|
| 1 | 0.15 M | 0.30 M | 46 |
| 2 | 0.20 M | 0.60 M | 61 |
| 3 | 0.20 M | 0.30 M | 61 |

(A)   rate = $k\,[N_2O_5]$

(B)   rate = $k\,[N_2O_5]^2$

(C)   rate = $k\,[N_2O_5]^{4/3}\,[O_2]^2$

(D)   rate = $k\,[N_2O_5]^2\,[O_2]$

*Knowledge Required:* (1) The meanings of the terms rate law, first order, and second order. (2) The algebraic form of the rate equation. (2) How to use the method of initial rates to determine the rate law.

*Thinking it Through:* The data represent cases where concentrations of reactants are varied and the corresponding rates of reactions are measured. Notice that experiments **1** and **3** represent cases where $[O_2]$ is held constant while $[N_2O_5]$ is varied. Experiments **2** and **3** represent cases where $[N_2O_5]$ is held constant and $[O_2]$ is varied. The next thing to notice is that the initial concentration of oxygen in experiment **2** is twice what it was in experiment **3**, yet the rate was the same in both experiments. This tells us that $[O_2]$ does not affect the rate of the reaction, eliminating choices **(C)** and **(D)**. The task now is to ascertain whether the reaction is first order in $[N_2O_5]$, choice **(A)**, or second order in $[N_2O_5]$, choice **(B)**. If the reaction is first order in $[N_2O_5]$, the increase in rate will be directly proportional to the increase in concentration. If the reaction is second order in $[N_2O_5]$, the rate will increase as the square of $[N_2O_5]$.

Checking first for a first order reaction by calculating the expected rate:

$$\frac{\text{rate}_2}{\text{rate}_1} = \frac{k[N_2O_5]_2}{k[N_2O_5]_1}$$

$$\text{rate}_2 = \frac{0.20\ \text{M}}{0.15\ \text{M}} \times 46 = 61$$

Since the predicted rate is that for a reaction that is first order in $N_2O_5$, choice **(A)** is confirmed as the correct answer.

**DY-4.** For the reaction

$$2H_2O_2 \rightarrow 2H_2O + O_2$$

which plot confirms that the rate is first order with respect to $H_2O_2$?

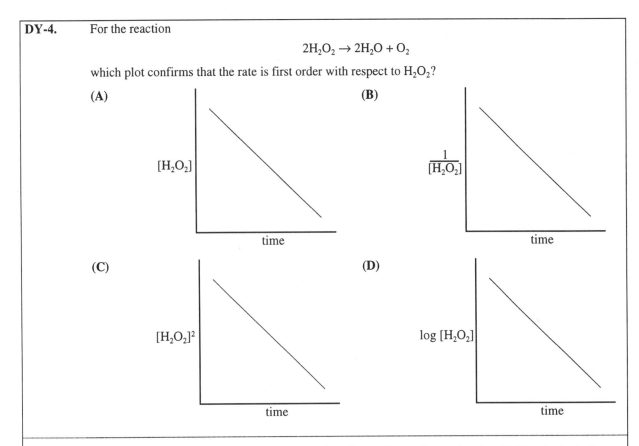

**(A)** [$H_2O_2$] vs time

**(B)** $\frac{1}{[H_2O_2]}$ vs time

**(C)** [$H_2O_2$]$^2$ vs time

**(D)** log [$H_2O_2$] vs time

*Knowledge Required:* (1) The meanings of the terms rate law, first order, and second order. (2) The algebraic form of the integrated rate equation. (2) How to use the graphical method to determine the rate law.

*Thinking it Through:* The integrated form of the first-order rate equation is $\ln\frac{[A]_0}{[A]} = kt$. Rearranging the equation in the form of a straight line ($y = mx + b$) yields

$$\ln [A] = -kt + \ln [A]_0$$

A plot of ln [A] (the concentration of hydrogen peroxide, which corresponds to $y$) as a function of $t$ (time, which corresponds to $x$) yields a straight line for a first order reaction. Of the choices offered, only **(D)** is possible, but there the *common* logarithm of concentration, rather than the *natural* logarithm is plotted against time. The result is the same, however. Since common logarithms and natural logarithms are directly proportional to each other, either will produce a straight line plot for a first order reaction, confirming choice **(D)** as the correct answer. Choice **(A)** results for zero order reactions. Choices **(B)** and **(C)** do not correspond to any reaction order.

Note: Choice **(B)** would correspond to the plot for a second-order reaction if the straight line had a positive slope.

**DY-5.** The activation energy for a particular reaction is 83.1 kJ·mol$^{-1}$. By what factor will the rate constant increase when the temperature is increased from 50.0 °C to 60.0 °C?

    **(A)**     10.0      **(B)**     2.53      **(C)**     0.927      **(D)**     0.395

*Knowledge Required:* (1) The meaning of activation energy and rate constant. (2) How to use the Arrhenius equation.

*Thinking it Through:* The equation that relates activation energy, rate constant, and temperature is the Arrhenius equation:

$$k = Ae^{-E_a/RT}$$

This equation is easier to use in the form

$$\ln k = \left(-\frac{E_a}{R}\right)\left(\frac{1}{T}\right) + \ln A$$

The problem asks for the ratio of $k_2/k_1$, which can be obtained by subtracting the equation at the first temperature, $T_1$, from that at the second temperature, $T_2$, and solving for $k_2/k_1$.

$$\ln k_2 - \ln k_1 = \left[\left(-\frac{E_a}{R}\right)\left(\frac{1}{T_2}\right) - \left(-\frac{E_a}{R}\right)\left(\frac{1}{T_1}\right)\right] + \ln A - \ln A$$

$$\ln \frac{k_2}{k_1} = \left(\frac{E_a}{R}\right)\left(\frac{1}{T_1} - \frac{1}{T_2}\right) = \left(\frac{83.1 \text{ kJ} \cdot \text{mol}^{-1}}{8.314 \text{ J} \cdot \text{K}^{-1}}\right)\left(\frac{1000 \text{ J}}{1 \text{ kJ}}\right)\left(\frac{1}{(273.15+50.0) \text{ K}} - \frac{1}{(273.15+60.0) \text{ K}}\right)$$

$$\ln \frac{k_2}{k_1} = 0.927$$

$$k_2/k_1 = e^{0.927} = 2.53$$

This is choice **(B)**, the correct answer. Choice **(C)** results from forgetting to take the antilog. Choice **(A)** reflects confusion with the notion that a 10 °C rise in temperature results in an approximate doubling of the reaction rate. Choice **(D)** is the ratio $k_1/k_2$.

**DY-6.** The half-life for the radioactive decay of $^{32}$P is 14.2 days. How many days would be required for a sample of a radiopharmaceutical containing $^{32}$P to decrease to 20% of its initial activity?

    **(A)**     33.0 d      **(B)**     49.2 d      **(C)**     71.0 d      **(D)**     286 d

*Knowledge Required:* (1) The knowledge that radioactive decay is a first-order process. (2) The integrated form of the first-order rate equation. (3) The relationship between half-life and rate constant.

*Thinking it Through:* For radioactive decay, $t_{1/2} = 0.693/k$, so for this reaction

$$k = \frac{0.693}{14.2 \text{ days}} = 0.0488 \text{ days}^{-1}$$

Now we can use the rate equation to find the time required to reach the ratio $[A]/[A]_0 = 0.20$.

$$\ln\left(\frac{[A]}{[A]_0}\right) = -kt$$

$$t = -\frac{\ln(0.20)}{0.0488 \text{ days}^{-1}} = 33.0 \text{ days}$$

The correct answer is choice **(A)**. Choice **(C)** is five half-lives, which would leave 1/32 of the original material. Choice **(D)** is twenty half-lives. Choice **(B)** is five half-lives multiplied by 0.693.

**DY-7.**   Consider the reaction

$$2NO_2(g) + F_2(g) \rightleftharpoons 2NO_2F(g)$$

A proposed mechanism for this reaction is

$$NO_2 + F_2 \rightleftharpoons NO_2F + F \quad \text{(slow)}$$
$$NO_2 + F \rightleftharpoons NO_2F \quad \text{(fast)}$$

What is the rate law for this mechanism?

(A)    rate $= k \dfrac{[NO_2F]^2}{[NO_2]^2 [F_2]}$

(B)    rate $= k [NO_2]^2 [F_2]$

(C)    rate $= k [NO_2] [F_2]$

(D)    rate $= k [NO_2] [F]$

*Knowledge Required:* (1) The fact that the slowest step in a reaction mechanism determines the rate law. (2) The algebraic form of the rate law expression.

*Thinking it Through:* The slow step is the reaction of $NO_2$ with $F_2$ to form $NO_2F$ and F. The rate equation is therefore expected to contain both $NO_2$ and $F_2$, with the concentration of each raised to the first power. That is choice (C). Choice (D) contains a product (here, a fluorine atom) of the reaction rather than only reactants. Choice (B) is a commonly chosen wrong answer, since it is derived from the stoichiometry of the overall reaction. Choice (A) is the equilibrium constant expression rather than the rate-law expression.

---

**DY-8.**   This mechanism has been proposed for the formation of ethylbenzene:

1    $CH_3CH_2Br + AlBr_3 \rightarrow AlBr_4^- + CH_3CH_2^+$

2    $CH_3CH_2^+ + C_6H_6 \rightarrow C_6H_6CH_2CH_3^+$

3    $C_6H_6CH_2CH_3^+ + AlBr_4^- \rightarrow AlBr_3 + HBr + C_6H_5CH_2CH_3$

Which substance serves as the catalyst?

(A)    $AlBr_3$

(B)    $AlBr_4^-$

(C)    $CH_3CH_2^+$

(D)    $C_6H_6CH_2CH_3^+$

*Knowledge Required:* (1) The role of a catalyst in chemical reactions.

*Thinking it Through:* Catalysts are substances that increase the rate of chemical reactions without themselves being consumed in the reaction. What we're looking for in the sequence of steps is a substance that is used in one step and regenerated in a later one. Step **1** uses $CH_3CH_2Br$ and $AlBr_3$. $AlBr_3$ is produced by step **3**, so there is no net loss of $AlBr_3$ in the overall reaction. It serves as the catalyst. Choice (A) is the correct answer. Every other choice is a reaction intermediate.

---

**DY-9.**   Consider this equilibrium:

$$2SO_2(g) + O_2(g) \underset{2}{\overset{1}{\rightleftharpoons}} 2SO_3(g)$$

The forward reaction (**1**) is proceeding at a certain rate at some temperature and pressure. When the pressure is increased, what might we expect for the forward reaction?

(A)    a greater rate of reaction and a greater yield of $SO_3$ at equilibrium

(B)    a greater rate of reaction and the same yield of $SO_3$ at equilibrium

(C)    a lesser rate of reaction and a smaller yield of $SO_3$ at equilibrium

(D)    a lesser rate of reaction and a greater yield of $SO_3$ at equilibrium

*Knowledge Required:* (1) How to use LeChatelier's principle. (2) The qualitative understanding of rate laws.

*Thinking it Through:* Considering LeChatelier's principle first, we see from the chemical equation that three moles of gas (2 mol $SO_2$ plus 1 mol of $O_2$) react to produce two moles of product. The equilibrium will shift to minimize the effect of the change, producing more $SO_3$. Choices (**B**) and (**C**) are therefore eliminated from consideration.

Increasing the pressure also increases the concentration of $SO_2$ and $O_2$, and one or the other (or both) are undoubtedly involved in the rate equation for this reaction. Increasing their concentrations is expected to increase the rate of reaction, leading to choice (**A**) as the correct answer. Choice (**D**) correctly recognizes the greater yield, but not the greater rate.

---

| **DY-10.** | In a chemical reaction involving the formation of an intermediate activated complex, which step must always be exothermic? |
|---|---|

| | |
|---|---|
| (**A**)    reactants → products | (**B**)    products → reactants |
| (**C**)    reactants → activated complex | (**D**)    activated complex → products |

*Knowledge Required:* (1) The general understanding of multi-step reaction mechanisms. (2) The general shape of energy *vs.* reaction coordinate diagrams. (3) The meaning of the terms *exothermic* and *endothermic*.

*Thinking it Through:* Spontaneous chemical reactions are usually exothermic, but not always. Choice (**A**) cannot be true. The reverse reaction of an exothermic reaction must be endothermic, so choice (**B**) is eliminated for the same reason as choice (**A**). The activated complex exists at the top of the energy barrier isolating reactants and products. Energetically speaking, it is downhill from the activated complex to either reactants or products. Choice (**D**), is the correct answer, is always exothermic. The remaining choice, (**C**), is always endothermic.

## Practice Questions

1.  The gas-phase reaction, $A_2 + B_2 \rightarrow 2AB$, proceeds by bimolecular collisions between $A_2$ and $B_2$ molecules. If the concentrations of both $A_2$ and $B_2$ are doubled, the reaction rate will change by a factor of

    (**A**)    1/2        (**B**)    $\sqrt{2}$        (**C**)    2        (**D**)    4

2.  For the reaction of chlorine and nitric oxide,

    $$2NO_{(g)} + Cl_{2(g)} \rightarrow 2NOCl_{(g)}$$

    doubling the concentration of chlorine doubles the rate of reaction. Doubling the concentration of both reactants increases the rate of reaction by a factor of eight. The reaction is

    (**A**)    first order in both NO and $Cl_2$.

    (**B**)    first order in NO and second order in $Cl_2$.

    (**C**)    second order in NO and first order in $Cl_2$.

    (**D**)    second order in both NO and $Cl_2$.

3.  Under certain conditions, the average rate of *appearance* of oxygen gas in the reaction

    $$2O_{3(g)} \rightarrow 3O_{2(g)}$$

    is $1.2 \times 10^{-3}$ atm·s$^{-1}$. What is the average rate, expressed in atm·s$^{-1}$, for the *disappearance* of $O_3$?

    (**A**)    $8.0 \times 10^{-4}$        (**B**)    $1.2 \times 10^{-3}$        (**C**)    $1.8 \times 10^{-3}$        (**D**)    $5.3 \times 10^{-3}$

**4.** Which reaction coordinate diagram represents a reaction in which the activation energy, $E_a$, is 50 kJ·mol$^{-1}$ and the $\Delta H_{rxn}$ is $-15$ kJ·mol$^{-1}$?

**(A)**

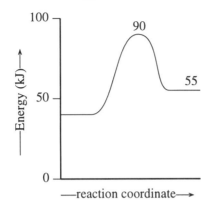

**(B)**

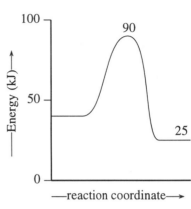

**(C)**

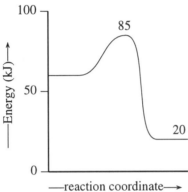

**(D)**

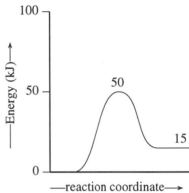

**5.** The rate law for the reaction

$$H_2O_2 + 2H^+ + 2I^- \rightarrow I_2 + 2H_2O$$

is rate = $k$ [H$_2$O$_2$] [I$^-$]. The overall order of the reaction is

**(A)** five.    **(B)** three.    **(C)** two.    **(D)** one.

**6.** The activity of a radioisotope is 3000 counts per minute at one time and 2736 counts per minute 48 hours later. What is the half-life of the radioisotope?

**(A)** 831 hr    **(B)** 521 hr    **(C)** 361 hr    **(D)** 1.44 hr

**7.** The half-life for the first order conversion of cyclobutane to ethylene,

$$C_4H_8(g) \rightarrow 2C_2H_4(g)$$

is 22.7 s at a particular temperature. How many seconds are needed for the partial pressure of cyclobutane to decrease from 100 mmHg to 10 mmHg?

**(A)** 52.0 s    **(B)** 75.4 s    **(C)** 90.0 s    **(D)** 227 s

**8.** A plot of reactant concentration as a function of time gives a straight line. What is the order of the reaction for this reactant?

**(A)** zero    **(B)** first    **(C)** second    **(D)** third

9. When the reaction

$$CH_3Cl_{(g)} + H_2O_{(g)} \rightarrow CH_3OH_{(g)} + HCl_{(g)}$$

was studied, the tabulated data were obtained. Based on these data, what are the reaction orders?

| | Initial Concentrations, M; Initial Rates, M·s⁻¹ | | |
|---|---|---|---|
| Exp | CH₃Cl | H₂O | Rate |
| 1 | 0.100 | 0.100 | 0.182 |
| 2 | 0.200 | 0.200 | 1.45 |
| 3 | 0.200 | 0.400 | 5.81 |

(A)     CH₃Cl: first order     H₂O: first order

(B)     CH₃Cl: first order     H₂O: second order

(C)     CH₃Cl: second order  H₂O: first order

(D)     CH₃Cl: second order  H₂O: second order

10. The reaction between acetone and bromine in acid solution is represented by the equation

$$CH_3OCH_{3(aq)} + Br_{2(aq)} + H_3O^+_{(aq)} \rightarrow products$$

The tabulated kinetic data were gathered. Based on these data, the experimental rate law is

| | Initial Concentrations, M; Initial Rates, M·s⁻¹ | | | |
|---|---|---|---|---|
| Exp | CH₃OCH₃ | Br₂ | H₃O⁺ | Rate |
| 1 | 0.30 | 0.050 | 0.050 | $5.8 \times 10^{-5}$ |
| 2 | 0.30 | 0.100 | 0.050 | $5.8 \times 10^{-5}$ |
| 3 | 0.30 | 0.050 | 0.100 | $1.2 \times 10^{-4}$ |
| 4 | 0.40 | 0.050 | 0.200 | $3.2 \times 10^{-4}$ |

(A)     rate = $k \, [CH_3OCH_3]^1 \, [Br_2]^1 \, [H_3O^+]^1$

(B)     rate = $k \, [CH_3OCH_3]^1 \, [Br_2]^0 \, [H_3O^+]^1$

(C)     rate = $k \, [CH_3OCH_3]^0 \, [Br_2]^0 \, [H_3O^+]^2$

(D)     rate = $k \, [CH_3OCH_3]^1 \, [Br_2]^1 \, [H_3O^+]^0$

11. All of these changes increase the value of the rate constant for a reaction *except*

(A)     decreasing the activation energy.

(B)     raising the temperature.

(C)     adding a catalyst.

(D)     increasing the concentration of reactants.

12. The experimental data from a certain reaction gives these three graphs. What is the most likely order for this reaction?

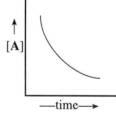 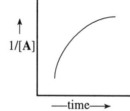 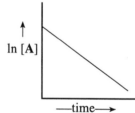

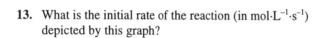

(A)     zero          (B)     first          (C)     second          (D)     third

13. What is the initial rate of the reaction (in mol·L⁻¹·s⁻¹) depicted by this graph?

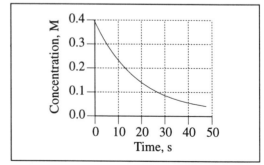

(A)     0.02          (B)     0.01          (C)     0.008          (D)     0.005

14. Which function of [**X**], plotted against time, will give a straight line for a second order reaction?

    (A)    [**X**]    (B)    [**X**]$^2$    (C)    ln [**X**]    (D)    1/[**X**]

15. The Arrhenius equation describes the relationship between the rate constant, $k$, and the energy of activation, $E_a$.

$$k = Ae^{-E_a/RT}$$

    In this equation, $A$ is an empirical constant, $R$ is the ideal-gas constant, $e$ is the base of natural logarithms, and $T$ is the absolute temperature. According to the Arrhenius equation,

    (A)    at constant temperature, reactions with lower activation energies proceed more rapidly.

    (B)    at constant temperature, reactions with lower activation energies proceed less rapidly.

    (C)    at constant energy of activation, reactions at lower temperatures proceed more rapidly.

    (D)    at constant energy of activation, reactions with smaller values of $A$ proceed more rapidly.

16. What will be the effect of increasing the temperature of reactants that are known to undergo an endothermic reaction?

    (A)    Both the rate of reaction and the value of the equilibrium constant increase.

    (B)    The rate of reaction increases and the value of the equilibrium constant decreases.

    (C)    The rate of reaction decreases and the value of the equilibrium constant increases.

    (D)    The rate of reaction increases and the value of the equilibrium constant is unchanged.

17. A change in temperature from 10 °C to 20 °C is found to double the rate of a particular chemical reaction. How did the change in temperature affect the reacting molecules?

    (A)    The average velocity of the molecules doubled.

    (B)    The average energy of the molecules doubled.

    (C)    The number of collisions per second doubled.

    (D)    The number of molecules above the reaction energy threshold doubled.

18. The value for the rate constant of a reaction can generally be expected to

    (A)    decrease with increasing temperature.

    (B)    increase with increasing temperature.

    (C)    decrease with increasing temperature only when the reaction is exothermic.

    (D)    increase with increasing temperature only when the reaction is exothermic.

19. Two reactions with different activation energies have the same rate at room temperature. Which statement correctly describes the rates of these two reactions at the same, higher temperature?

    (A)    The reaction with the larger activation energy will be faster.

    (B)    The reaction with the smaller activation energy will be faster.

    (C)    The two reactions will continue to occur at the same rates.

    (D)    A prediction cannot be made without additional information.

20. A catalyst increases the rate of a reaction by

    (A)    changing the mechanism of the reaction.

    (B)    increasing the activation energy of the reaction.

    (C)    increasing the concentration of one or more of the products.

    (D)    decreasing the difference in relative energy of the reactants and products.

21. Consider the reaction

$$Cl_2(aq) + H_2S(aq) \rightarrow S(s) + 2H^+(aq) + 2Cl^-(aq)$$

The rate equation for this reaction is

$$rate = k\,[Cl_2]\,[H_2S]$$

Which of these mechanisms is (or are) consistent with this rate equation?

    I    $Cl_2 + H_2S \rightarrow H^+ + Cl^- + Cl^+ + HS^-$         (slow)

         $Cl^+ + HS^- \rightarrow H^+ + Cl^- + S$         (fast)

    II   $H_2S \rightleftharpoons H^+ + HS^-$         (fast equilibrium)

         $Cl_2 + HS^- \rightarrow 2Cl^- + H^+ + S$         (slow)

    (A)    I only                         (B)    II only

    (C)    Both I and II                  (D)    Neither I or II

22. A certain reaction has a $\Delta H = -75$ kJ and an activation energy of 40 kJ. A catalyst is found that lowers the activation energy of the forward reaction by 15 kJ. What is the activation energy of the reverse reaction in the presence of this same catalyst?

    (A)    25 kJ        (B)    60 kJ        (C)    90 kJ        (D)    100 kJ

23. What is the relationship between the equilibrium constant ($K_c$) of a reaction and the rate constants for the forward ($k_f$) and reverse ($k_r$) reactions?

    (A)    $K_c = k_f k_r$                     (B)    $K_c = k_f/k_r$

    (C)    $K_c = 1/(k_f k_r)$                 (D)    $K_c = k_f - k_r$

24. If the half-life of a reaction is independent of concentration, the reaction can be
    I  first order         II  second order         III  zero order

    (A)    I and II only.                 (B)    II and III only.

    (C)    I only.                        (D)    II only.

25. Initial rate data for the reaction

    $$2H_2(g) + Cl_2(g) \rightarrow 2HCl(g)$$

    are given in the table. What is the rate law for this reaction?

| Experiment | $[H_2]_{initial}$ | $[Cl_2]_{initial}$ | Rate in $M \cdot s^{-1}$ |
|---|---|---|---|
| 1 | 0.0020 M | 0.0050 M | $2.5 \times 10^{-3}$ |
| 2 | 0.0020 M | 0.0025 M | $1.3 \times 10^{-3}$ |
| 3 | 0.0015 M | 0.0025 M | $1.3 \times 10^{-3}$ |
| 4 | 0.0050 M | 0.0010 M | $0.5 \times 10^{-3}$ |

    (A)    rate = $k\,[Cl_2]^2$                 (B)    rate = $k\,[Cl_2]$

    (C)    rate = $k\,[H_2]$                   (D)    rate = $k\,[H_2]\,[Cl_2]$

26. The rate of conversion from cyclopropane to propene in the gas phase at constant temperature is a first order reaction. From the information in the graph, which statement about the half-life of the conversion reaction is true?

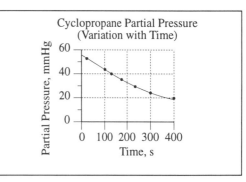

(A)    It is approximately equal to 150 s.    (B)    It is approximately equal to 270 s.

(C)    It is approximately equal to 420 s.    (D)    It cannot be determined.

27. Consider this reaction.

$$4NH_3(g) + 3O_2(g) \rightleftharpoons 2N_2(g) + 6H_2O(l)$$

If the rate of formation of $N_2$ is 0.10 M·s$^{-1}$, what is the corresponding rate of disappearance of $O_2$?

(A)    1.5 M·s$^{-1}$    (B)    0.30 M·s$^{-1}$

(C)    0.15 M·s$^{-1}$    (D)    0.10 M·s$^{-1}$

28. Which line segment represents the activation energy for the reaction between **C** and **D** to form **A** and **B**?

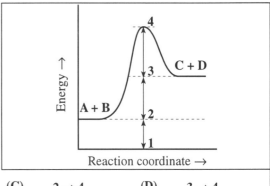

(A)    $1 \rightarrow 4$    (B)    $2 \rightarrow 3$    (C)    $2 \rightarrow 4$    (D)    $3 \rightarrow 4$

29. The decomposition of hydrogen peroxide in the presence of iodide ion is believed to occur via this mechanism.

$$H_2O_2(aq) + I^-(aq) \rightarrow H_2O(l) + IO^-(aq)$$

$$H_2O_2(aq) + IO^-(aq) \rightarrow H_2O(l) + O_2(g) + I^-(aq)$$

In this mechanism, $I^-(aq)$ is

(A)    a catalyst.    (B)    a reactant in the overall reaction.

(C)    the activated complex.    (D)    a product of the overall reaction.

30. Which statement most accurately describes the behavior of a catalyst?

(A)    A catalyst increases the $\Delta G$ of a reaction and hence the forward rate.

(B)    A catalyst reduces the $\Delta H$ of a reaction and hence the temperature needed to produce products.

(C)    A catalyst reduces the activation energy for a reaction and increases the rate of a reaction.

(D)    A catalyst increases the equilibrium constant and final product concentrations.

## Answers to Study Questions

| | | | | | |
|---|---|---|---|---|---|
| 1. | B | 5. | B | 9. | A |
| 2. | B | 6. | A | 10. | D |
| 3. | A | 7. | C | | |
| 4. | D | 8. | A | | |

## Answers to Practice Questions

| | | | | | |
|---|---|---|---|---|---|
| 1. | D | 11. | D | 21. | A |
| 2. | C | 12. | B | 22. | D |
| 3. | A | 13. | A | 23. | B |
| 4. | B | 14. | D | 24. | C |
| 5. | C | 15. | A | 25. | B |
| 6. | C | 16. | A | 26. | B |
| 7. | B | 17. | D | 27. | C |
| 8. | A | 18. | B | 28. | D |
| 9. | B | 19. | A | 29. | A |
| 10. | B | 20. | A | 30. | C |

# Equilibrium

When doing stoichiometric calculations, the assumption often made is that the reaction goes to completion. This is a convenient assumption when focusing on calculations involving mole ratios and limiting reagents, but there are many examples of commercially and biologically important chemical reactions that do *not* go to completion. Rather, appreciable amounts of reactants and products remain in the reaction mixture once equilibrium is reached. When viewed macroscopically, the concentrations of all reactants and products remain constant, but not necessarily equal, over time.

What is happening at the molecular level is far from static. To emphasize this, the term *dynamic equilibrium* is often used. The rate at which reactants form products is balanced by the rate of the reverse reaction, although this statement does not imply equal *amounts* of both reactants and products, only that the *rates* are the same.

In this part of your study of general chemistry, it is typical to explore how a chemical system comes to equilibrium and how that equilibrium is represented. Interpretation of equilibrium constants, calculation of concentrations in equilibrium systems, equilibrium in gas phase reactions, relationship of enthalpy to equilibrium, acid–base equilibria, and solubility equilibria all fall into this area of study.

## Study Questions

**EQ-1.** When the reversible reaction $N_2(g) + O_2(g) \rightleftharpoons 2NO(g)$ has reached a state of equilibrium,

    **(A)**     no further reaction occurs.

    **(B)**     the total moles of products must equal the remaining moles of reactant.

    **(C)**     the addition of a catalyst will cause formation of more NO.

    **(D)**     the concentration of each substance in the system will be constant.

*Knowledge Required:* (1) The nature of a reversible reaction. (2) Understanding of the equilibrium state.

*Thinking it Through:* When a reaction has reached equilibrium, it does not mean that all chemical activity has stopped. Rather, at equilibrium, the macroscopic view indicates constant (but seldom equal) concentrations for each substance, making Choice **(D)** the correct response. Choice **(A)** is a commonly held misconception, one that you will not choose if you remember the concept of *dynamic* equilibrium. It is also untrue that the total moles of products must equal the remaining moles of reactant, choice **(B)**. The relative amounts of material present at equilibrium will depend greatly on the position of the equilibrium, revealed in quantitative problems by the value of the equilibrium constant. Choice **(C)** is based on another common misconception about equilibrium reactions. Addition of a catalyst, while it may increase the *rate* at which equilibrium is achieved, does not affect the *position* of equilibrium.

*Note:* The reversible reaction of nitrogen and oxygen to form nitrogen monoxide is an important one in the control of emissions from combustion engines, for NO serves as the trigger for the formation of photochemical smog.

**EQ-2.** Xenon tetrafluoride, $XeF_4$, can be prepared by heating Xe and $F_2$ together as represented by this equation.

$$Xe(g) + 2F_2(g) \rightleftharpoons XeF_4(g)$$

What is the equilibrium constant expression for this reaction?

    **(A)**    $K_c = \dfrac{[XeF_4]}{[Xe][F_2]}$          **(B)**    $K_c = \dfrac{[XeF_4]}{2[Xe][F_2]}$

    **(C)**    $K_c = \dfrac{[XeF_4]}{[Xe][F_2]^2}$          **(D)**    $K_c = \dfrac{[Xe][F_2]}{[XeF_4]}$

*Knowledge Required:* (1) Algebraic form of the equilibrium expression.

*Thinking it Through:* For the general chemical equation $aA + bB \rightleftharpoons cC + dD$ representing a reversible reaction, and with concentrations expressed in moles per liter, the equilibrium expression, $K_c$, is written as

$$K_c = \frac{[C]^c [D]^d}{[A]^a [B]^b}$$

This expression is sometimes called the *Law of Mass Action*. Note that the equilibrium constant, $K_c$, always places products in the numerator of the expression and reactants in the denominator; this criterion alone eliminates choice (**D**). The second feature to note is that the *coefficients* in the chemical equation become *exponents* in the equilibrium expression. The only coefficient here is the 2 in front of the reactant $F_2$ molecule, and that becomes the exponent shown in choice (**C**), the correct choice. Choice (**A**) does not include the coefficient, and choice (**B**) uses the coefficient as a multiplier rather than as an exponent.

Several different subscripts are used with the equilibrium constant. When all reactants or products that appear in the equilibrium expression are gases, and pressures are used instead of molarity, the symbol for the equilibrium constant is $K_P$. If weak acids and bases are under investigation, the equilibrium constant symbol is written as $K_a$ or $K_b$. When the equilibrium constant refers to the autoionization of water, the symbol $K_w$ is used. Solubility product equilibria express the ion concentrations of partially soluble salts. The equilibrium constant symbol $K_{sp}$ is used for solubility product equilibria.

---

**EQ-3.**     In pure water at 60 °C, these relationships hold.

$$[H_3O^+] = [OH^-] = 3.1 \times 10^{-7} \text{ M}$$

It is reported that an aqueous solution at 60 °C has $[H_3O^+]$ equal to $1.0 \times 10^{-7}$ M. Such a solution is

(**A**)     neutral.     (**B**)     basic.     (**C**)     acidic.     (**D**)     impossible.

---

*Knowledge Required:* (1) The meaning of the ion-product constant, $K_w$, for water. (2) Interpretation of $K_w$ at conditions other than 25 °C.

*Thinking it Through:* The ion-product constant for water, also called the dissociation constant for water, refers to the autoionization of water. This is the equilibrium reaction that defines the pH scale at 25 °C.

$$2H_2O(l) \rightleftharpoons H_3O^+(aq) + OH^-(aq)$$

$$K_w = [H_3O^+][OH^-] = [1.0 \times 10^{-7}][1.0 \times 10^{-7}] = 1.0 \times 10^{-14}$$

At 60 °C, $K_w = [H_3O^+][OH^-] = [3.1 \times 10^{-7}][3.1 \times 10^{-7}] = 9.6 \times 10^{-14}$

We are used to having the concentrations of the $H_3O^+(aq)$ ion and the $OH^-(aq)$ ion being expressed at 25 °C, but this fundamental relationship is true at any temperature.

If $[H_3O^+] > [OH^-]$, the solution is acidic.

If $[H_3O^+] = [OH^-]$, the solution is neutral.

If $[H_3O^+] < [OH^-]$, the solution is basic.

Autoionization is an endothermic reaction, so the reaction at higher temperatures must shift to the right, and that shift is reflected in the larger value for $K_w$ at 60 °C. If $[H_3O^+]$ is $1.0 \times 10^{-7}$ M at 60 °C, and $[OH^-] = K_w/[H_3O^+]$, then $[OH^-] = (9.6 \times 10^{-14})/(1.0 \times 10^{-7} \text{ M}) = 9.6 \times 10^{-7}$ M. Hydroxide ion concentration is almost ten times greater than hydronium ion concentration. The solution is basic, which is choice (**B**). Choice (**A**) may be selected if you do not understand that the position of equilibrium will shift with increasing temperature. Choice (**C**) does not seem plausible unless you mistakenly think the equilibrium shifts in the opposite direction at higher temperatures. Choice (**D**), and responses such as this, are usually (but not always) incorrect. Infrequently a test writer may want to determine if a student can recognize when additional data will be needed for solving a problem, but experienced test writers generally avoid writing such questions.

**EQ-4.** A mixture of 2.0 mol of $CO_{(g)}$ and 2.0 mol of $H_2O_{(g)}$ was allowed to come to equilibrium in a 10.0-L flask at a high temperature. If $K_c = 4.0$, what is the molar concentration of $H_{2(g)}$ in the equilibrium mixture? The equation for the reaction is this.

$$CO_{(g)} + H_2O_{(g)} \rightleftharpoons CO_{2(g)} + H_{2(g)}$$

**(A)** 0.67 M **(B)** 0.40 M **(C)** 0.20 M **(D)** 0.13 M

***Knowledge Required:*** (1) Algebraic form of the equilibrium expression. (2) Substitutions into the equilibrium expression to calculate equilibrium concentrations. (3) Alternative approaches to solving equilibrium problems.

***Thinking it Through:*** In this gas-phase reaction, carbon monoxide and water vapor are allowed to come to equilibrium, forming carbon dioxide and hydrogen. The equilibrium expression for the reaction is this.

$$K_c = \frac{[CO_2][H_2]}{[CO][H_2O]}$$

Substituting into this equilibrium expression, and letting $x$ represent the molarity of carbon dioxide that forms, $x$ will also represent the molarity of hydrogen gas that forms because the mole ratio in the chemical equation is 1:1. This same variable $x$ will represent the molarity of carbon monoxide that has *reacted* and the molarity of water vapor that has *reacted*; all mole ratios in the balanced chemical equation are 1:1. Subtracting this value from the original 2.0 mol will give the concentration at equilibrium for both CO and $H_2O$. The reaction takes place in a 10.0-L container, so the initial concentrations of $CO_{(g)}$ and $H_2O_{(g)}$ are each 0.20 M. $K_c$ can now be expressed showing these relationships.

$$K_c = 4.0 = \frac{(x)(x)}{(0.20-x)(0.20-x)} = \frac{x^2}{(0.20-x)^2}$$

Now $x$ can be found by solving the quadratic equation. It is much quicker to simplify the algebra, in this particular case, by taking the square root of each side.

$$\sqrt{4.0} = \sqrt{\frac{x^2}{(0.20-x)^2}} \quad \text{and} \quad 2.0 = \frac{x}{0.20-x}$$

Then solve for $x$: $2.0 \times (0.20 - x) = x$, so $0.40 - (2.0 \times x) = x$, and $0.40 = 3.0 \times (x)$. This makes $x = 0.13$ M. This is the correct choice **(D)**. Be sure to do a quick check by substituting the value for $x$ into the original equilibrium expression. The calculated $K_c$ should agree reasonably well with the value given in the statement of the problem.

$$K_c = \frac{[0.13][0.13]}{[0.20-0.13][0.20-0.13]} = 3.4 \approx 4.0$$

Another way to check the possible responses without really solving the problem is to substitute each response value into the equilibrium expression. Then evaluate the expression to see if it yields the value of $K_c$. Only one will be near the correct response. Do not expect exact agreement even then. Rounding errors, particularly when subtracting, can affect the calculation. Remember that negative concentrations, although mathematically possible, cannot occur in the real chemical world.

Considering the other responses that were offered, choice **(B)** might have been selected if you thought all four concentrations had to be equal. Choice **(A)** is the difference between 0.20 and 0.133. This is the molar concentration of each *reactant* at equilibrium, not that of hydrogen, which is a product. Choice **(C)** would be the concentration of hydrogen if the reaction went to completion.

**Note:** Think about simplifying approaches to each question and choose the one that works for you. The faculty who write ACS Exams know you have a limited amount of time and are seeking to help you display your knowledge in the time available.

**EQ-5.** When a sample of $NO_2$ is placed in a container, this equilibrium is rapidly established.

$$2NO_{2(g)} \rightleftharpoons N_2O_{4(g)}$$

If this equilibrium mixture is a darker color at high temperatures or at low pressures, which statement about the reaction is true?

(A) The reaction is exothermic and $NO_2$ is darker in color than $N_2O_4$.

(B) The reaction is exothermic and $N_2O_4$ is darker in color than $NO_2$.

(C) The reaction is endothermic and $NO_2$ is darker in color than $N_2O_4$.

(D) The reaction is endothermic and $N_2O_4$ is darker in color than $NO_2$.

*Knowledge Required:* (1) How temperature and pressure affect chemical equilibrium (LeChatelier's Principle). (2) Qualitative relationship between enthalpy of reaction and point of equilibrium.

*Thinking it Through:* Initially, there may not seem to be enough information to answer this question. The enthalpy of the reaction has not been given for the reaction as written, and there is no statement about whether the reaction is endothermic or exothermic. The key to understanding this equilibrium mixture is to note that "high temperatures" and "low pressures" both give the *same* macroscopic effect; the mixture becomes *darker*. Notice that there are two moles of gas on the reactant side and only one mole of gas on the product side. If the pressure in the reaction vessel is lowered, the equilibrium position must shift to favor the side with the *larger number of moles*, which is the left side for this reaction. $NO_2$ must be the dark-colored gas. This observation eliminates choices (B) and (D), which identify the darker-colored gas as $N_2O_4$. Deciding between (A) and (C) rests on the observation that the equilibrium mixture becomes *darker at high temperatures*, meaning the reaction is exothermic in the forward direction. Increasing the temperature forces the equilibrium to the left. Choice (A) is correct.

**EQ-6.** Consider this reaction at equilibrium.

$$2SO_{2(g)} + O_{2(g)} \rightleftharpoons 2SO_{3(g)} \qquad \Delta H = -198 \text{ kJ}$$

Which of these changes would cause an increase in the $SO_3 / SO_2$ mole ratio?

(A) adding a catalyst

(B) removing $O_{2(g)}$

(C) decreasing the temperature

(D) decreasing the pressure

*Knowledge Required:* (1) How temperature and pressure affect chemical equilibrium (LeChatelier's Principle). (2) How the sign of the value for enthalpy change indicates heat flow in a chemical reaction. (3) Qualitative relationship between enthalpy of reaction and point of equilibrium.

*Thinking it Through:* In this case, the enthalpy change for the reaction is negative, meaning that the reaction as written is **exothermic**. If the goal is to increase the $SO_3 / SO_2$ mole ratio, anything that shifts the position of the equilibrium to the right will be of help. Adding a catalyst, choice (A), only affects the *rate* at which equilibrium is achieved, not the concentrations of reactants and products. A catalyst has no effect on the mole ratio. Removing oxygen gas, choice (B), drives the reaction to the left, so it affects the mole ratio of $SO_3 / SO_2$ in the wrong direction. Decreasing the temperature, choice (C), drives the reaction to the right, which increases the ratio of $SO_3 / SO_2$; this is the correct response. In an exothermic reaction, think of the heat given off as a product. The equation itself may be written to emphasize this idea.

$$2SO_{2(g)} + O_{2(g)} \rightleftharpoons 2SO_{3(g)} + \text{heat}$$

If the temperature of the reaction is decreased, the effect is similar to that observed when removing a product from the equilibrium mixture. The reaction will proceed to the right, increasing the concentration of $SO_3$ relative to that of $SO_2$. Decreasing the pressure *decreases* the concentration of $SO_3$ relative to that of $SO_2$. Observe that there are three moles of gas on the left side of the equation as written and two moles on the right hand side. Decreasing the pressure, choice (D), shifts the equilibrium to the left.

| EQ-7. | The solubility of solid silver chromate, $Ag_2CrO_4$, which has $K_{sp}$ equal to $9.0 \times 10^{-12}$ at 25 °C, is determined in water and in two different aqueous solutions. Predict the relative solubility of $Ag_2CrO_4$ in the three solutions. | List of Solutions Used |
|---|---|---|

| | List of Solutions Used |
|---|---|
| I. | pure water |
| II. | 0.1 M $AgNO_3$ |
| III. | 0.1 M $Na_2CrO_4$ |

(A)  I = II = III

(B)  I < II < III

(C)  II = III < I

(D)  II < III < I

**Knowledge Required:** (1) The interpretation of solubility-product constants, $K_{sp}$. (2) Common-ion effect. (3) Qualitative understanding of solubility equilibria.

**Thinking it Through:** When partially soluble crystalline solids dissolve in pure water to form an aqueous solution, a dynamic equilibrium is established between the solid that separates into hydrated ions and the reforming of the solid phase from the hydrated ions. For silver chromate, this is the equilibrium reaction.

$$Ag_2CrO_4(s) \rightleftharpoons 2Ag^+(aq) + CrO_4^{2-}(aq) \qquad \text{Equation 1}$$

The equilibrium constant, $K_{sp}$, is also known as the solubility product constant or simply the solubility product. Since the solid silver chromate does not affect the equilibrium position, it is omitted from the $K_{sp}$ expression. The ion concentrations are those found in a saturated solution. This is the solubility product expression for $Ag_2CrO_4$.

$$K_{sp} = [Ag^+]^2 [CrO_4^{2-}] = 9.0 \times 10^{-12} \qquad \text{Equation 2}$$

Notice that the value of this solubility product constant is small; there are relatively few silver or chromate ions in aqueous solution at 25 °C. If changing the solution composition shifts the equilibrium in Equation 1 to the right, more solid will dissociate and the solubility will increase relative to the solubility of silver chromate in pure water at 25 °C. If the equilibrium shifts to the left as different solution compositions are used, then the solubility relative to that in pure water will decrease.

Consider first the $AgNO_3$ solution. 0.1 M $Ag^+(aq)$ ions are already present in the solution, which would inhibit the addition of $Ag^+(aq)$ from dissolving $Ag_2CrO_4(s)$. The solubility equilibrium reaction is pushed to the left. Using a 0.1 M $Na_2CrO_4$, solution has a similar effect. With 0.1 M $CrO_4^{2-}(aq)$ ions already present in the solution, the addition of chromate ions is suppressed. Both solutions illustrate the "common ion" effect, an often-used experimental method for precipitating, or "salting out," the maximum amount of a precipitate from solution. As it is now established that both solutions decrease the solubility of the silver chromate, choices (A) and (B) can be eliminated. In focusing on the remaining choices, note that both the $AgNO_3$ and the $Na_2CrO_4$ solutions are 0.1 M solutions, so they each contribute 0.1 M of a common ion, either silver ion or chromate ion. Before picking choice (C), however, remember that the equilibrium expression shows that the concentration of the silver ion is *squared*, giving silver ion concentration in the dissolving solution greater effect on the position of equilibrium. Silver chromate is least soluble in 0.1 M silver nitrate solution, followed by 0.1 M sodium chromate, and then pure water, choice (D).

| EQ-8. | Typical "hard" water contains about $2.0 \times 10^{-3}$ mol of $Ca^{2+}$ per liter. Calculate the maximum concentration of fluoride ion that could be present in hard water. Assume fluoride is the only anion present that will precipitate calcium ion. | | |
|---|---|---|---|

| Substance | $K_{sp}$, 25 °C |
|---|---|
| $CaF_2(s)$ | $4.0 \times 10^{-11}$ |

(A)  $2.0 \times 10^{-3}$ M  (B)  $1.4 \times 10^{-4}$ M  (C)  $1.0 \times 10^{-5}$ M  (D)  $2.0 \times 10^{-8}$ M

**Knowledge Required:** (1) The interpretation of solubility equilibrium constants, $K_{sp}$. (2) Quantitatively using solubility equilibrium to calculate ion concentration.

**Thinking it Through:** For calcium fluoride, this is the equilibrium reaction.

$$CaF_2(s) \rightleftharpoons Ca^{2+}(aq) + 2F^-(aq)$$

The formula shows there are two fluoride ions per formula unit. That value affects the $K_{sp}$ expression in two ways. Dissolving produces two fluoride ions for every calcium ion, so the concentration of fluoride ions from dissolving $CaF_2$ will always be twice that of calcium ions. The statement of the problem does not, however, suggest that the calcium ion and fluoride ion present in the solution come *only* from dissolving calcium fluoride. The problem asks *what concentration of fluoride ion can be present*, regardless of where either ion comes from.

$$K_{sp} = 4.0 \times 10^{-11} = [Ca^{2+}] [F^-]^2$$

Substituting the given values and solving for the fluoride ion concentration is the next step. The volume is one liter, so the number of moles can be used as well as using the molarity.

$$4.0 \times 10^{-11} = [2.0 \times 10^{-3}] [F^-]^2$$

$$[F^-]^2 = \frac{4.0 \times 10^{-11}}{2.0 \times 10^{-3}}$$

Solving for $[F^-]$ gives a value of $1.4 \times 10^{-4}$ M, which is correct choice (**B**). Forgetting to take the square root of $[F^-]^2$ leads to incorrect choice (**D**). Choice (**A**) fails to use the equilibrium constant expression at all, and assumes that calcium ions and fluoride ions are present in equimolar concentrations. Choice (**C**) results from squaring the *calcium* ion concentration before dividing it into the value for the solubility-product constant.

---

**EQ-9.** At 298 K, the equilibrium constant for this reaction

$$H_2(g) + \tfrac{1}{2}O_2(g) \rightleftharpoons H_2O(l)$$

| Compound and State | $\Delta G_f^\circ$, kJ·mol$^{-1}$ |
|---|---|
| $H_2O(l)$ | −237 |
| $H_2O(g)$ | −229 |

(**A**) is larger than the $K_{eq}$ for $H_2(g) + \tfrac{1}{2}O_2(g) \rightleftharpoons H_2O(g)$

(**B**) has a value of 1.0 at equilibrium.

(**C**) can not be computed since data on $O_2$ and $H_2$ are not provided.

(**D**) will have the same value as the $K_{eq}$ for $H_2(g) + \tfrac{1}{2}O_2(g) \rightleftharpoons H_2O(g)$

**Knowledge Required:** (1) The relationship of the change in Gibbs Free Energy, $\Delta G_f^\circ$, and the equilibrium constant. (2) Interpretation of $\Delta G_f^\circ$ data to determine relative values of equilibrium constants.

**Thinking it Through:** The mathematical relationship between $\Delta G^\circ$ for a reaction and the equilibrium constant is given here. $R$ is the gas constant in units of J·K$^{-1}$mol$^{-1}$ and $T$ is the Kelvin temperature.

$$\Delta G^\circ = -RT \ln (K_{eq})$$

For this problem, it is not necessary to solve the free energy equation, only to determine qualitative relationships. The negative sign means that the larger the negative value for $\Delta G^\circ$, the larger the positive value for $\ln(K_{eq})$ and thus for $K_{eq}$ itself. Therefore, the value of $\Delta G_f^\circ$ for liquid water, being more negative that $\Delta G_f^\circ$ for gaseous water, indicates a larger equilibrium constant for the formation of liquid water from its elements in their standard states. This is the correct choice (**A**). Choice (**D**) is eliminated because the equilibrium constant for the formation of liquid water and gaseous water cannot be the same at the same temperature since the values of $\Delta G_f^\circ$ differ. Choice (**C**) is eliminated because the standard free energy for any element in its standard state is defined as equal to zero. Choice (**B**) expresses a misconception that all equilibrium constants for spontaneous reactions are equal to 1.0.

| EQ-10. | Given these acids and their $K_a$ values, which sequence lists the corresponding anions in order of *increasing* base strength? | | Equilibrium Values, $K_a$ | |
|---|---|---|---|---|
| | | | $HC_2H_3O_2$ | $1.8 \times 10^{-5}$ |
| | | | HCN | $6.2 \times 10^{-10}$ |
| | | | HF | $7.2 \times 10^{-4}$ |

(A) $CN^-, C_2H_3O_2^-, F^-$  (B) $C_2H_3O_2^-, CN^-, F^-$

(C) $F^-, C_2H_3O_2^-, CN^-$  (D) $F^-, CN^-, C_2H_3O_2^-$

***Knowledge Required:*** (1) The interpretation of acid equilibrium constants, $K_a$. (2) Meaning of terms associated with weak and strong acids and bases. (3) Brønsted–Lowry acid-base definition.

***Thinking it Through:*** Strong acids are completely ionized in water, so equilibrium constant expressions are not used for them. A weak acid is only partially ionized. The value of the equilibrium constant for a weak acid gives information about the degree of ionization.

$$HA(aq) + H_2O(l) \rightleftharpoons H_3O^+(aq) + A^-(aq)$$

acid    base    conjugate acid    conjugate base

The equilibrium constant, $K_a$, is defined to include the concentration of water, which does not change within the limits of error for weak acid–base equilibria in dilute solutions. Since the constant concentration of water molecules does not affect the point of equilibrium, water is omitted from the expression.

$$K_a = \frac{[H_3O^+][A^-]}{[HA]}$$

A small numerical value for $K_a$ means that there are very few ions formed relative to the number of molecules present. Of the three acids in the problem, HCN is by far the weakest, followed by $HC_2H_3O_2$, and then HF. The anions that form in the ionization are themselves conjugate Brønsted–Lowry bases of the respective acid. The strongest base will be best at reacting with water and pick up a proton to reform the acid. The strength of these anions as bases are *in reverse order* of the strength of each weak acid. The general equation is

$$A^-(aq) + H_2O(l) \rightleftharpoons HA(aq) + OH^-(aq)$$

base    acid    conjugate acid    conjugate base

Choice (C) correctly lists the anions in their order of increasing base strength, based on decreasing acid strength. The only other choice that represents a common error is choice (A), which might indicate that you were choosing based on the acid strength, not the base strength. Both choices (B) and (D) are ineffective in this question, because they are based on nothing except possibly the order (or reverse order) of the acids as listed in the table of equilibrium values.

## Practice Questions

1. Which is a proper description of chemical equilibrium?

(A) The frequencies of reactant and of product collisions are identical.

(B) The concentrations of products and reactants are identical.

(C) The velocities of product and reactant molecules are identical.

(D) Reactant molecules react to form products as fast as product molecules are reacting to form reactants.

2. Consider this reaction.

$$P_4(s) + 6Cl_2(g) \rightleftharpoons 4PCl_3(g)$$

What is the correct $K_c$ expression for this reaction?

(A) $\quad K_c = \dfrac{1}{[Cl_2]^6}$

(B) $\quad K_c = \dfrac{[PCl_3]^4}{[P_4][Cl_2]^6}$

(C) $\quad K_c = \dfrac{[PCl_3]^4}{[Cl_2]^6}$

(D) $\quad K_c = \dfrac{[P_4][Cl_2]^6}{[PCl_3]^4}$

3. Consider this reaction.

$$2SO_3(g) \rightleftharpoons 2SO_2(g) + O_2(g)$$

What is the correct $K_p$ expression for this reaction?

(A) $\quad K_p = \dfrac{P_{SO_2}^2 P_{O_2}}{P_{SO_3}^2}$

(B) $\quad K_p = \dfrac{P_{SO_2} P_{O_2}}{P_{SO_3}}$

(C) $\quad K_p = \dfrac{(2P_{SO_2})^2 P_{O_2}}{(2P_{SO_3})^2}$

(D) $\quad K_p = \dfrac{2P_{SO_2}^2 P_{O_2}}{2P_{SO_3}^2}$

4. Consider this reaction.

$$2C(s) + O_2(g) \rightleftharpoons 2CO(g)$$

What is the equilibrium expression for this reaction?

(A) $\quad K_c = \dfrac{[CO]}{[C][O_2]}$

(B) $\quad K_c = \dfrac{[CO]^2}{[C]^2[O_2]}$

(C) $\quad K_c = \dfrac{[2CO]}{[2C][O_2]}$

(D) $\quad K_c = \dfrac{[CO]^2}{[O_2]}$

5. Carbon monoxide gas reacts with hydrogen gas at elevated temperatures to form methanol according to this equation.

$$CO(g) + 2H_2(g) \rightleftharpoons CH_3OH(g)$$

When 0.40 mol of CO and 0.30 mol of $H_2$ are allowed to reach equilibrium in a 1.0 L container, 0.060 mol of $CH_3OH$ are formed. What is the value of $K_c$?

(A)    0.50          (B)    0.98          (C)    1.7          (D)    5.4

6. The graph shows the variation of concentration with time for this reaction at 25 °C.

$$3A(aq) \rightleftharpoons B(aq) + 2C(aq)$$

What is the value of the equilibrium constant at time $t_2$?

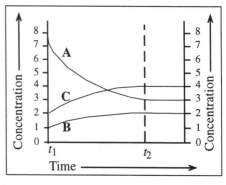

(A)    1.2          (B)    0.84          (C)    0.57          (D)    0.22

7. The value of an equilibrium constant can be used to predict each of these *except* the

    (A)    direction of a reaction.

    (B)    extent of a reaction.

    (C)    quantity of reactant(s) remaining at equilibrium.

    (D)    time required to reach equilibrium.

8. Chemical equilibrium is the result of

    (A)    formation of products equal in mass to the mass of the reactants.

    (B)    the unavailability of one of reactants.

    (C)    a stoppage of further reaction.

    (D)    opposing reactions attaining equal speeds.

9. Consider this gas-phase reaction.

$$H_2(g) + I_2(g) \rightleftharpoons 2HI(g) \quad \Delta H = 53 \text{ kJ}$$

Which reaction characteristics will be affected by a change in temperature?

    **1.** value of equilibrium constant        **2.** equilibrium concentrations

    (A)    **1** only        (B)    **2** only        (C)    **1** and **2**        (D)    neither **1** nor **2**

10. Which factors will affect both the position of equilibrium and the value of the equilibrium constant for this reaction?

$$N_2(g) + 3H_2(g) \rightleftharpoons 2NH_3(g) \quad \Delta H = -92 \text{ kJ}$$

    (A)    increasing the volume of the container

    (B)    adding more nitrogen gas

    (C)    removing ammonia gas

    (D)    lowering the temperature

11. The reversible reaction

$$2H_2(g) + CO(g) \rightleftharpoons CH_3OH(g) + \text{heat}$$

is carried out by mixing carbon monoxide and hydrogen gases in a closed vessel under high pressure with a suitable catalyst. After equilibrium is established at high temperature and pressure, all three substances are present. If the pressure on the system is lowered, with the temperature kept constant, what will be the result?

    (A)    The amount of $CH_3OH$ will be increased.

    (B)    The amount of $CH_3OH$ will be decreased.

    (C)    The amount of each substance will be unchanged.

    (D)    The amount of each substance will be increased.

12. Consider this reaction, carried out at constant volume.

$$2SO_2(g) + O_2(g) \rightleftharpoons 2SO_3(g) \qquad \Delta H = -198 \text{ kJ}$$

The concentration of $O_2(g)$ at equilibrium increases if

    (A)    $SO_2$ is added to the system.        (B)    $SO_3$ is added to the system.

    (C)    the temperature of the system is lowered.    (D)    an inert gas is added to the system.

13. Consider this reaction, carried out at constant temperature and volume.

$$2SO_2(g) + O_2(g) \rightleftharpoons 2SO_3(g)$$

What is the effect of removing some $SO_3$ from a system initially at equilibrium?

   (A)   The concentration of $SO_2$ decreases more than the concentration of $O_2$.

   (B)   The concentration of $SO_2$ increases more than the concentration of $O_2$.

   (C)   The concentration of $SO_2$ and $O_2$ remain the same.

   (D)   The concentration of $SO_2$ and $O_2$ decrease equally.

14. Consider this reaction, carried out at constant temperature and volume.

$$PCl_5(g) \rightleftharpoons PCl_3(g) + Cl_2(g)$$

How can the position of equilibrium for this reaction be shifted to the right?

   (A)   addition of a catalyst

   (B)   removal of $Cl_2$

   (C)   addition of an inert gas at constant volume

   (D)   removal of $PCl_5$

15. Consider this reaction.

$$\text{heat} + CaSO_3(s) \rightleftharpoons CaO(s) + SO_2(g)$$

What change will cause an increase in the pressure of $SO_2(g)$ when equilibrium is re-established?

   (A)   increasing the reaction temperature    (B)   decreasing the volume of the container

   (C)   adding some more $CaSO_3$               (D)   removing some of the $CaO(s)$

16. In which reaction would an increase in pressure at constant temperature have no effect on the relative amounts of the substances present in the equilibrium mixture? All substances are gases.

   (A)   $2NO + O_2 \rightleftharpoons 2NO_2 + \text{heat}$    (B)   $\text{heat} + N_2 + O_2 \rightleftharpoons 2NO$

   (C)   $N_2 + 3H_2 \rightleftharpoons 2NH_3 + \text{heat}$    (D)   $2CO + O_2 \rightleftharpoons 2CO_2 + \text{heat}$

17. The numerical value of the equilibrium constant for any chemical change is affected by changing the

   (A)   nature of the catalyst.          (B)   concentration of the products.

   (C)   the pressure.                    (D)   the temperature.

18. Solid carbon and gaseous carbon dioxide are placed in a closed vessel and allowed to come to equilibrium at a high temperature according to this equation.

$$\text{heat} + C(s) + CO_2(g) \rightleftharpoons 2CO(g)$$

If the pressure on the system is increased and the temperature kept constant, what will be the result?

   (A)   The amount of CO will increase and that of C and $CO_2$ will decrease.

   (B)   The amount of CO will decrease and that of C and $CO_2$ will increase.

   (C)   The amount of each of the three substances will increase.

   (D)   The amount of each of the three substances will decrease.

19. Consider this reaction.

$$AB_3(g) \rightleftharpoons A(g) + 3B(g)$$

What is the equilibrium constant expression if the initial concentration of $AB_3$ is 0.1 M and the equilibrium concentration of $A$ is represented by $x$? Assume the initial concentrations of $A$ and $B$ are both zero.

(A) $\dfrac{x - 3x}{0.1 - x}$

(B) $\dfrac{x - x^3}{(0.1 - x)^3}$

(C) $\dfrac{x \cdot x^3}{(0.1 - 3x)^3}$

(D) $\dfrac{x \cdot (3x)^3}{0.1 - x}$

20. At 0 °C the ion product constant of water, $K_w$, is $1.2 \times 10^{-15}$. The pH of pure water at this temperature is

(A) 6.88     (B) 7.00     (C) 7.46     (D) 7.56

21. What is the correct equation for the ion product constant of water at 25 °C?

(A) $[H_3O^+][OH^-] = 10^{-14}$

(B) $[H_3O^+] + [OH^-] = 10^{-14}$

(C) $\dfrac{[H_3O^+]}{[OH^-]} = 10^{-14}$

(D) $\dfrac{[H_3O^+][OH^-]}{[H_2O]} = 10^{-14}$

22. What is the solubility product, $K_{sp}$, of $Mg(OH)_2$ if its solubility in water is $1.6 \times 10^{-4}$ mol·L$^{-1}$?

(A) $1.6 \times 10^{-11}$     (B) $2.6 \times 10^{-8}$     (C) $3.2 \times 10^{-4}$     (D) $4.1 \times 10^{-12}$

23. A saturated solution of $MgF_2$ contains $1.16 \times 10^{-3}$ mol of $MgF_2$ per liter at a certain temperature. What is the $K_{sp}$ of $MgF_2$ at this temperature?

(A) $2.7 \times 10^{-6}$     (B) $1.6 \times 10^{-8}$     (C) $3.1 \times 10^{-9}$     (D) $6.2 \times 10^{-9}$

24. What will be the result if 100 mL of 0.06 M $Mg(NO_3)_2$ is added to 50 mL of 0.06 M $Na_2C_2O_4$? Assume the reaction is taking place at 25 °C.

| Substance | $K_{sp}$, 25 °C |
|-----------|-----------------|
| $MgC_2O_4(s)$ | $8.6 \times 10^{-5}$ |

(A) No precipitate will form.

(B) A precipitate will form and an excess of $Mg^{2+}$ ions will remain in solution.

(C) A precipitate will form and an excess of $C_2O_4^{2-}$ ions will remain in solution.

(D) A precipitate will form but neither ion is present in excess.

25. The addition of solid $Na_2SO_4$ to an aqueous solution in equilibrium with solid $BaSO_4$ will cause

| Substance | $K_{sp}$, 25 °C |
|-----------|-----------------|
| $BaSO_4(s)$ | $1.5 \times 10^{-9}$ |

(A) no change in $[Ba^{2+}]$ in solution.

(B) more $BaSO_4$ to dissolve.

(C) precipitation of more $BaSO_4$.

(D) an increase in the $K_{sp}$ of $BaSO_4$.

26. In a 0.050 M solution of a weak monoprotic acid, $[H^+] = 1.8 \times 10^{-3}$. What is its $K_a$?

(A) $3.6 \times 10^{-2}$     (B) $9.0 \times 10^{-5}$     (C) $6.7 \times 10^{-5}$     (D) $1.6 \times 10^{-7}$

**27.** A solution of sodium acetate in water is observed to become more alkaline as the temperature is raised. Which conclusion can be drawn? This is the equation for the reaction.

$$Na^+(aq) + C_2H_3O_2^-(aq) + H_2O(l) \rightleftharpoons HC_2H_3O_2(aq) + Na^+(aq) + OH^-(aq)$$

**(A)** The forward reaction proceeds with evolution of heat.

**(B)** The forward reaction proceeds with absorption of heat.

**(C)** Acetic acid is less soluble in hot water than in cold water.

**(D)** At higher temperatures, $Na^+(aq) + OH^-(aq) \rightarrow NaOH(aq)$ will occur.

**28.** A 0.15 M solution of a weak acid is found to be 1.3% ionized. What is its $K_a$?

**(A)** $1.3 \times 10^{-2}$   **(B)** $2.0 \times 10^{-3}$   **(C)** $1.1 \times 10^{-3}$   **(D)** $2.6 \times 10^{-5}$

**29.** What is the value of the equilibrium constant $K$, for a reaction for which $\Delta G°$ is equal to –5.20 kJ at 50 °C?

**(A)** 0.144   **(B)** 0.287   **(C)** 6.93   **(D)** 86.4

**30.** For the reaction

$$NH_4Cl(s) \rightarrow NH_3(g) + HCl(g)$$

$\Delta H° = +176$ kJ and $\Delta G° = +91.2$ kJ at 298 K. What is the value of $\Delta G$ at 1000 K?

**(A)** –109 kJ   **(B)** –64 kJ   **(C)** +64 kJ   **(D)** +109 kJ

# Answers to Study Questions

| | | | |
|---|---|---|---|
| 1. D | 5. A | 9. A | |
| 2. C | 6. C | 10. C | |
| 3. B | 7. D | | |
| 4. D | 8. B | | |

# Answers to Practice Questions

| | | |
|---|---|---|
| 1. D | 11. B | 21. A |
| 2. C | 12. B | 22. A |
| 3. A | 13. A | 23. D |
| 4. D | 14. B | 24. B |
| 5. D | 15. A | 25. C |
| 6. A | 16. B | 26. C |
| 7. D | 17. D | 27. B |
| 8. D | 18. B | 28. D |
| 9. C | 19. D | 29. C |
| 10. D | 20. C | 30. A |

# Electrochemistry and Redox

Electrochemistry is an important area of study for its applications to everyday life. Products such as voltaic cells and batteries and processes such as corrosion are examples of common electrochemical processes. In electrochemistry, there is interchange of chemical and electrical energy either through spontaneous chemical change or by adding energy to force a nonspontaneous change to take place. Qualitative predictions and quantitative calculations of cell potential are included in this unit. Fundamental to understanding the topics of electrochemistry are the ability to describe the movement of electrons during chemical reactions. This is tracked by assigning oxidation states and identifying chemical species responsible for observed changes. The Nernst equation· provides a way for understanding the relationship between cell potential and the concentration of the cell components. The stoichiometry of electrolytic processes and the relationship of cell potential to free energy are also considered in studying electrochemistry.

## Study Questions

---

**ER-1.** What is the oxidation number of phosphorus in $H_3PO_2$?

    **(A)**    +1          **(B)**    +2          **(C)**    +3          **(D)**    +4

---

*Knowledge Required:* (1) The rules for assigning oxidation numbers.

*Thinking it Through:* Oxidation numbers are always assigned for *each* atom, not for groups of atoms. For a monoatomic ion, the oxidation number is equal to the charge on the ion. In covalent compounds, oxidation numbers are assigned to atoms even though they do not have an actual charge. The oxidation number assignment is simply a method of keeping track of electron movement in chemical reactions. The most common oxidation number for hydrogen is +1 in its compounds, and the most common oxidation number for oxygen is –2. These values can be predicted from each element's position on the periodic table. Hydrogen is in Group IA; it tends to lose one electron in forming compounds. Oxygen is in Group VIA, and usually forms bonds in which two additional electrons are shared or transferred to the oxygen, giving it the stable octet of valence electrons. The oxidation number of phosphorus is not as easy to predict, but in this neutral compound it is the value needed for all the oxidation numbers to sum to zero. (If the species being asked about had been a polyatomic ion, the sum of oxidation numbers would have added up to be the charge on the ion.) $H_3PO_2$ is neutral, so $3(+1) + x + 2(-2) = 0$. Solving this equation: $x = +1$, which is choice **(A)**.

**Note:** The term *oxidation number* is also referred to as *oxidation state* because an oxidation number represents an oxidation state of an element.

---

**ER-2.** Nickel is a transition element and has a variable valence. Using a nickel salt, 2.00 faradays plate out 39.2 g of nickel. What form of nickel ion is in the solution of this salt?

| Data | |
|---|---|
| Atomic Molar Mass, Ni | 58.7 g·mol$^{-1}$ |
| 1 faraday ($F$) | 96,485 C |

    **(A)**    $Ni^+$          **(B)**    $Ni^{2+}$          **(C)**    $Ni^{3+}$          **(D)**    $Ni^{4+}$

---

*Knowledge Required:* (1) The stoichiometry of electrolytic processes. (2) The relationships among coulombs, faradays, and moles of electrons.

*Thinking it Through:* If 39.2 g of nickel plate out, this corresponds to 0.668 mol of nickel, or roughly two-thirds of a mole of nickel. If 2.0 $F$ were used to plate out 0.6678 mol of nickel, then 3.00 $F$ (3.00 mol of electrons) will be required to plate out 1.00 mol of nickel. This means the reaction for the reduction must be $Ni^{3+}(aq) + 3e^- \rightarrow$ $Ni_{(s)}$, making choice **(C)** the correct response. Choice **(B)** may be incorrectly chosen if the fact that 2 faradays are used is misinterpreted. $Ni^{2+}$ is also the most common oxidation state for Ni, also making this incorrect response very attractive. Choices **(A)** and **(D)** represent oxidation states that do not commonly exist for nickel.

**ER-3.** Consider this reaction.

$$2Fe^{3+}(aq) + 2I^-(aq) \rightarrow 2Fe^{2+}(aq) + I_2(aq)$$

Which statement is true for the reaction?

(A)   $Fe^{3+}$ is oxidized.

(B)   $Fe^{3+}$ increases in oxidation number.

(C)   $Fe^{3+}$ is reduced.

(D)   $I^-$ is reduced.

*Knowledge Required:* (1) The definitions of oxidation and reduction. (2) Interpretation of changes in oxidation number from a balanced chemical equation.

*Thinking it Through:* In any oxidation–reduction or redox reaction, at least one substance must be losing electrons and one substance must be gaining those electrons; both processes are occurring simultaneously. Changes in oxidation state will be evident in the equation when comparing the oxidation numbers of the reactants to the products. $Fe^{3+}$ changes to $Fe^{2+}$. This can only happen through the *gain* of electrons, which defines the process of *reduction*; the *oxidation number decreases*. The correct choice is (C). The accompanying change is the loss of electrons from the iodide ion in forming iodine, $I_2$. The *loss* of electrons defines the process of *oxidation*; the *oxidation number increases* from negative one to zero for each mole of iodide ion. Be very sure that you have these definitions straight, because analysis of more complex equations will depend on these ideas. The basic vocabulary of redox reactions is summarized in this diagram.

$$2\ (Fe^{3+} + e^- \longrightarrow Fe^{2+})$$

process of reduction; gain of $e^-$

$$2Fe^{3+}(aq) \quad + \quad 2I^-(aq) \quad \rightarrow \quad 2Fe^{2+}(aq) \quad + \quad I_2(aq)$$

process of oxidation; loss of $e^-$

$$2\ I^- \longrightarrow I_2 + 2e^-$$

Discarding the other choices is a matter of comparing the statements to the definitions. Choice (A) is not correct because $Fe^{3+}$ did not *lose* electrons to form $Fe^{2+}$; it *gained* electrons. Choice (B) is incorrect for a similar reason; the oxidation number did not increase, it *decreased* from +3 to +2. Choice (D) is incorrect because $I^-$ is not reduced, it is *oxidized*, and loses electrons in this reaction.

**ER-4.** What time is required to plate 2.08 g of copper at a constant current flow of 1.26 A?

$$Cu^{2+}(aq) + 2e^- \rightarrow Cu(s)$$

| Pertinent Data | |
| --- | --- |
| Atomic Molar Mass, Cu | 63.5 $g \cdot mol^{-1}$ |
| 1 faraday (F) | 96,485 C |

(A)   41.8 min   (B)   83.6 min   (C)   133 min   (D)   5016 min

*Knowledge Required:* (1) The stoichiometry of electrolytic processes. (2) The relationships among amperes, faradays, time, and moles of electrons.

*Thinking it Through:* An ampere (A) is a coulomb (C) of electric charge passing a point each second. A mole of electrons carries a charge of 96,485 C or 1 faraday (F). The time required to plate a given amount of copper depends on the number of moles of copper present and its ionic charge. In this case, two moles of electrons are required to change every mole of copper(II) ion to copper metal. This is the relationship to solve the problem. Pay special attention to the cancellation of units, which can help you reason through a problem in electrochemistry.

$$2.08g\ Cu \times \frac{1\ mol\ Cu}{63.5\ g\ Cu} \times \frac{2\ mol\ electrons}{1\ mol\ Cu} \times \frac{96,485\ C}{1\ mol\ electrons} \times \frac{1\ s}{1.26\ C} \times \frac{1\ min}{60\ s} = 83.6\ min$$

This is correct choice (B). Incorrect choice (A) fails to take into account the fact that *two* moles of electrons are required to make the conversion from $Cu^{2+}$ to Cu. Choice (C) has incorrectly multiplied by 1.26 A rather than dividing. Choice (D) is the number of seconds required, not minutes.

**ER-5.** Which statement about this redox reaction is correct?

$$2MnO_4^-(aq) + 5H_2O_2(aq) + 6H^+(aq) \rightarrow 2Mn^{2+}(aq) + 5O_2(g) + 8H_2O(g)$$

**(A)** $O_2$ acts as the oxidant in this reaction.

**(B)** $H_2O_2$ acts as a reducing agent.

**(C)** $H_2O_2$ acts as an oxidizing agent.

**(D)** Only oxidation has taken place.

*Knowledge Required:* (1) The definitions of oxidizing agent and reducing agent. (2) Interpretation of changes in oxidation number from a balanced chemical equation.

*Thinking it Through:* You can recognize this reaction as a redox reaction by a quick check of oxidation numbers. Manganese in $MnO_4^-$ has an oxidation number of +7. (The algebra is $x + 4(-2) = -1$; $x = +7$.) On the right hand side of the equation, $Mn^{2+}$ is produced. This means that each manganese atom has gained five electrons, this is the process of *reduction*. If manganese in $MnO_4^-$ is reduced, it has enabled the oxidation of some other species in the equation, and is called the *oxidizing agent* or the *oxidant*. Checking the other reactant, $H_2O_2$, we find that one of the products formed is free oxygen gas, with an assigned oxidation state of zero, as is the case for all free elements. Hydrogen peroxide does not contain oxygen in its usual oxidation state of -2. Assuming that each hydrogen retains its usual +1 oxidation state, then $2(+1) + 2(y) = 0$ and $y = -1$ per oxygen atom in hydrogen peroxide. You may need this information for other problems, but it is not essential for this problem. It is essential to note that the only way that oxygen can change from a negative oxidation state to zero is by *releasing* electrons, which is the process of *oxidation*. $H_2O_2$ enables the reduction of $MnO_4^-$ in this reaction, making $H_2O_2$ the *reducing agent* or *reductant*. This is correct choice **(B)**, and eliminates choice **(C)**. Choice **(A)** identifies one of the products, but redox vocabulary only labels the reactants. The products are used to identify changes in oxidation state. Choice **(D)** violates the basic tenet of redox reactions. Both oxidation and reduction must happen together.

**Note:** Hydrogen peroxide is one of a number of substances that can act as either an oxidizing agent or a reducing agent, depending on the relative tendency of other reactants to gain or lose electrons. Here $H_2O_2$ is being used to as a reducing agent that has the effect of decolorizing. The $MnO_4^-$ ion is an intense purple, but the $Mn^{2+}$ ion is a very pale pink and appears colorless at low concentrations.

**ER-6.** An oxidation-reduction reaction in which 3 electrons are transferred has $\Delta G° = +18.55$ kJ at 25 °C. What is the value of $E°$?

| Data | | |
|---|---|---|
| 1 F | = | 96,485 C·mol$^{-1}$ |
| 1 V | = | 1 J·C$^{-1}$ |

**(A)** +0.192 V    **(B)** −0.064 V    **(C)** −0.192 V    **(D)** −0.577 V

*Knowledge Required:* (1) The relationship of free energy to cell potential. (2) Meanings of units.

*Thinking it Through:* Standard free energy is related to standard cell potential, as shown by this equation.

$$\Delta G° = -nFE°$$

The free energy change is a function of the number of moles of electrons exchanged, the value of the faraday constant, and the standard cell potential. In this case, the value of the faraday constant and the relationship between volts and joules per coulomb are both given in the problem. In some cases, these constants will be located together in a table of useful information rather than specifically given in the question itself.

The value given for $\Delta G°$ is positive, an indication that the reaction is not spontaneous. This means the value of $E°$ will be negative, also indicative of a nonspontaneous cell. This eliminates choice **(A)** from further consideration. To choose among the negative cell potentials, the value of $E°$ can be calculated by substituting into the free energy and cell potential relationship.

$$+18.55 \text{ kJ} = -3 \text{ mol electrons} \times \frac{96,485 \text{ C}}{\text{mol of electrons}} \times E°$$

$$E° = 6.41 \times 10^{-5} \frac{\text{kJ}}{\text{C}} \times \frac{1000 \text{ J}}{1 \text{ kJ}} = 0.0642 \text{ V}$$

Choice **(B)** is correct. Choice **(C)** results from omitting the number of moles of electrons in the calculation. Choice **(D)** results from multiplying by the number of moles of electrons rather than dividing.

| ER-7. | Which reaction will occur if each substance is in its standard state? Assume potentials are given in water at 25 °C. | **Standard Reduction Potentials** | $E°$ |
|---|---|---|---|
| | | $Ni^{2+}(aq) + 2e^- \rightarrow Ni(s)$ | –0.28 V |
| | | $Sn^{4+}(aq) + 2e^- \rightarrow Sn^{2+}(s)$ | +0.15 V |
| | | $Br_2(l) + 2e^- \rightarrow 2Br^-(aq)$ | +1.06 V |

**(A)**  $Ni^{2+}$ will oxidize $Sn^{2+}$ to give $Sn^{4+}$

**(B)**  $Sn^{4+}$ will oxidize $Br^-$ to give $Br_2$

**(C)**  $Br_2$ will oxidize $Ni(s)$ to give $Ni^{2+}$

**(D)**  $Ni^{2+}$ will oxidize $Br_2$ to give $Br^-$

**Knowledge Required:** (1) The meaning of standard reduction potentials. (2) The use of standard reduction potentials to predict chemical change. (3) The relationship of standard reduction potentials to the Activity Series.

**Thinking it Through:** Standard reduction potentials are convenient values for predicting whether or not a redox reaction will take place. If a half-reaction has a high positive value of $E°$, it will be most likely to be reduced and therefore act as the best oxidizing agent. In this sequence of three half-reactions, the change from $Br_2(l)$ to $2Br^-(aq)$ with the gain of 2 electrons has the highest positive potential. It is only listed in choice **(C)**, so consider that change first. It is true that changing from nickel solid to nickel ion is a process of oxidation, so this is promising. While it is not *necessary* in this case, it is useful and will be essential to calculate the full cell potential in other cases. It is done here to show the process.

$$Br_2(l) + 2e^- \rightarrow 2Br^-(aq) \qquad +1.06 \text{ V}$$
$$\underline{Ni(s) \rightarrow Ni^{2+}(aq) + 2e^- \qquad +0.28 \text{ V}}$$
$$Br_2(l) + Ni(s) \rightarrow Ni^{2+}(aq) + 2Br^-(aq) \qquad +1.34 \text{ V}$$

This high positive cell potential is associated with a spontaneous reaction that will take place, so choice **(C)** still is the correct response. Be sure to check that $Br_2$ is indeed reduced, and can serve as the oxidizing agent to change nickel solid to nickel ion. Note that if you reverse the order of the chemical half-reaction, as was done with the nickel half reaction, the sign of the potential is also reversed. The general relationship is this.

$$E^o_{reduction} = -E^o_{oxidation}$$

Checking the other possible choices, **(A)** combining cell potentials for $Ni^{2+}$ as an oxidizing agent with that of $Sn^{2+}$ will give this result.

$$Ni^{2+}(aq) + 2e^- \rightarrow Ni(s) \qquad -0.28 \text{ V}$$
$$\underline{Sn^{2+}(s) \rightarrow Sn^{4+}(aq) \qquad -0.15 \text{ V}}$$
$$Ni^{2+}(aq) + Sn(s) \rightarrow Ni(s) + Sn^{4+}(aq) \qquad -0.43 \text{ V}$$

This is a negative cell potential and confirms the original observation that $Ni^{2+}(aq)$ is the *poorest* oxidizing agent on this particular short list. Checking choice **(B)** is not necessary if you observe that it will be intermediate in cell potential between choices **(A)** and **(C)**. (It does give a negative cell potential as well.) The reaction in choice **(D)** is impossible for both changes suggested, $Ni^{2+}$ to $Ni(s)$ and $Br_2(l)$ to $Br^-$ are *reductions*.

**Note:** Try to avoid learning some arbitrary rule such as the best oxidizing agent is on the bottom left. That will not serve you well if the list of potentials is presented with the most positive first, which is the usual convention. Standard reduction potentials also can be combined to calculate cell potentials. Cell potentials can then be related to the position of equilibrium for redox reactions, a skill we will practice shortly.

| ER-8. | Calculate the standard cell potential, $E°$, for this reaction. | **Standard Reduction Potentials** | $E°$ |
|---|---|---|---|
| | $Br_2(l) + 2Ce^{3+}(aq) \rightarrow 2Br^-(aq) + 2Ce^{4+}(aq)$ | $Ce^{4+}(aq) + e^- \rightarrow Ce^{3+}(aq)$ | +1.61 V |
| | | $Br_2(l) + 2e^- \rightarrow 2Br^-(aq)$ | +1.06 V |

**(A)**  –2.67 V   **(B)**  –2.16 V   **(C)**  –0.55 V   **(D)**  +2.67 V

**Knowledge Required:** (1) The meaning of standard reduction potentials. (2) How to combine standard reduction potentials to predict cell potential for a chemical reaction.

**Thinking it Through:** Standard cell potentials for redox reactions can be calculated from the proper combination of half-cell standard reduction potentials. Looking at the target equation, the liquid bromine in changing to aqueous bromide ion through the process of reduction. The reduction potential for that step will be +1.06 V. The accompanying change is from cerium in the +3 oxidation state to cerium in the +4 oxidation state; this is the process of oxidation. The potential for the oxidation step will be –1.61 V. Reversing the direction of electron flow from the listed reduction potential reverses the sign, as discussed in study problem **ER-7** above.

Combining these two half-reactions gives the cell potential for the reaction.

$$Br_2(l) + 2e^- \rightarrow 2Br^-(aq) \qquad +1.06 \text{ V}$$
$$\underline{2Ce^{3+}(aq) \rightarrow 2Ce^{4+}(aq) + 2e^- \qquad -1.61 \text{ V}}$$
$$2Ce^{3+}(aq) + Br_2(l) \rightarrow 2Br^-(aq) + 2Ce^{4+}(aq) \qquad -0.55 \text{ V} \quad \text{This is choice (C).}$$

The most common misconception in combining cell potentials is to think that when the coefficients change for a half-reaction, the potential changes. If you observed that *two* $Ce^{3+}$ changed to *two* $Ce^{4+}$ ions with the loss of *two* electrons, you may have the idea that the potential should be doubled as well. This assumption will lead to incorrect choice (B). Although this process is essential for functions such as free energy changes that depend on the amount of material present, it is not true for standard reduction potentials. Cell potentials are directly proportional to the free energy change *per electron*. The potential is independent of the number of moles present. Another possible error is simply to add the two half-cell potentials, failing to analyze the problem for oxidation and reduction changes. This will lead to either incorrect choices (A) or (D), depending on what sign assumption is made.

---

**ER-9.** During the electrolysis of a dilute aqueous solution of $K_2SO_4$, the solution around the cathode

(A) becomes slightly acidic.

(B) becomes slightly alkaline.

(C) remains neutral.

(D) becomes more dilute.

*Knowledge Required:* (1) The electrolysis of aqueous solutions. (2) Definitions of anode and cathode in electrochemical cells.

*Thinking it Through:* The cells considered so far in this study guide have been spontaneous or galvanic cells. This questions deals with an opposite process, electrolysis, in which electrical work is added to the system to cause a nonspontaneous redox reaction to take place. Overall, if a direct electric current is passed through water, in the presence of enough ions to carry the current, water will undergo a nonspontaneous redox reaction to produce hydrogen and oxygen gases. As long as the ions dissolved do not undergo oxidation or reduction more easily than water itself, any dilute aqueous solution will have the same electrolysis reaction as water itself. $K^+$, being the ion of an active metal, will not easily pick up an electron to form $K(s)$. Hydrogen from water is far more easily reduced. The sulfate ion, containing sulfur in its highest oxidation state of +6, cannot be oxidized further. It is energetically easier to oxidize the oxygen from water to oxygen gas. This is the overall electrolysis reaction.

$$2H_2O(l) \xrightarrow{\text{direct current, H}^+} 2H_2(g) + O_2(g)$$

The redox changes that take place can be identified in the usual manner. Note that the sum of the two half reactions is equivalent to the overall chemical reaction just given.

$$4H_2O + 4e^- \longrightarrow 2H_2 + 4OH^-$$
process of reduction; gain of $e^-$

$$2H_2O(l) \rightarrow 2H_2(g) + O_2(g)$$

process of oxidation; loss of $e^-$
$$2H_2O \longrightarrow O_2 + 4H^+ + 4e^-$$

In any cell, the process of **oxidation** takes place at the **anode** and **reduction** takes place at the **cathode**. Notice that in the half reaction of reduction, hydroxide ions are a product, accounting for the observation that the solution around the cathode becomes more alkaline, correct choice (B). If the anode and cathode were reversed, incorrect choice (**A**) might be picked, because it *is* true around the anode. Choice (**C**), while it is accurate for the solution as a whole, is not true around the cathode, given the half-reactions that are taking place. Choice (**D**) is untrue, for no additional water molecules are being produced to dilute the nonreactive solute, $K_2SO_4$. If anything, the salt concentration will be increasing as water is being consumed in the electrolysis reaction to produce hydrogen and oxygen gases.

**Note:** You may have seen electrolysis demonstrated in class or shown on a videodisc or on a CD-Rom. This is often used to show the 2:1 mole ratio of hydrogen gas to oxygen gas that is produced.

---

**ER-10.** Consider this reaction.

$$Cu^{2+}(aq) + Fe(s) \rightarrow Cu(s) + Fe^{2+}(aq \qquad E° = 0.78 \text{ V}$$

What is the value of $E$ when $[Cu^{2+}]$ is equal to 0.040 M and $[Fe^{2+}]$ is equal to 0.40 M?

| The Nernst Equation |
| --- |
| $E = E^o - \dfrac{0.0592}{n}\log(Q)$ |

(**A**)    0.72 V          (**B**)    0.75 V          (**C**)    0.81 V          (**D**)    0.84 V

---

*Knowledge Required:* (1) Use of the Nernst Equation to calculate cell potential under nonstandard conditions.

*Thinking it Through:* The Nernst equation expresses the dependence of cell potential on concentration of the cell components. The equation itself may be given in the problem, or may be found in a general table of useful equations in other ACS exams. There may also be questions where you are expected to know the relationship, or derive it from changes in free energy. The value of $Q$ is defined in the same manner as an equilibrium constant, but remember that these cell concentrations are not standard state concentrations.. For this problem, $Q$ is evaluated by this expression.

$$Q = \frac{[Fe^{2+}]}{[Cu^{2+}]} = \frac{0.40 \text{ M}}{0.040 \text{ M}} = 10$$

Before substituting this value into the Nernst equation and calculating the cell potential, it is worthwhile to predict how this $Q$ value will affect the potential of the cell. The given voltage at standard conditions is 0.78 V, and this term will be modified by the expression $-\dfrac{0.0592}{n}\log Q$ with $Q$ being equal to 10 and $n$, the number of moles of electrons being exchanged, equal to 2. The log of 10 is 1, so the standard cell potential will be *decreased* by 0.0592 divided by 2. This means the cell potential must be *less than* the standard cell potential, eliminating choices (**C**) and (**D**). A quick calculation using the Nernst Equation confirms that the answer is choice (**B**), since the cell potential is reduced by 0.03 V.

$$E = 0.78 \text{ V} - \frac{0.0592}{2}\log(10) = 0.78 \text{ V} - 0.0296 \text{ V} = 0.78 \text{ V} - 0.03 \text{ V} = 0.75 \text{ V}$$

You might choose (**A**) in error if you neglected to divide by $n = 2$. Both (**C**) and (**D**) are the result of adding, rather than subtracting the nonstandard cell potential term, $-\dfrac{0.0592}{n}\log Q$.

## Practice Questions

1. In which pair does the named element have the same oxidation number?

   **(A)** sulfur in $H_2S_2O_7$ and in $H_2SO_4$     **(B)** mercury in $Hg^{2+}$ and in $Hg_2^{2+}$

   **(C)** oxygen in $Na_2O_2$ and in $H_2O$     **(D)** cobalt in $Co(NH_3)_6^{3+}$ and in $Co(NO_3)_2$

2. In which compound does vanadium have the *lowest* oxidation state?

   **(A)** $V_2O_5$     **(B)** $V_2O_3$     **(C)** $VO_2$     **(D)** $VO$

3. What is the oxidation number of chromium in $Na_2Cr_2O_7$?

   **(A)** +12     **(B)** +6     **(C)** +3     **(D)** −2

4. Which of these elements can display the largest number of different oxidation states?

   **(A)** aluminum     **(B)** magnesium     **(C)** manganese     **(D)** mercury

5. Which statement is true for this reaction?

   $$Fe(s) + Cu^{2+}(aq) \rightarrow Cu(s) + Fe^{2+}(aq)$$

   **(A)** $Cu^{2+}$ is oxidized.     **(B)** $Cu^{2+}$ gains in oxidation number.

   **(C)** $Cu^{2+}$ is reduced.     **(D)** Fe is reduced.

6. Which change requires an oxidizing agent to produce the indicated product?

   **(A)** $S_2O_3^{2-} \rightarrow S_4O_6^{2-}$     **(B)** $Zn^{2+} \rightarrow Zn$

   **(C)** $ClO^- \rightarrow Cl^-$     **(D)** $SO_3 \rightarrow SO_4^{2-}$

7. Which statement is true for this reaction?

   $$Zn(s) + CuSO_4(aq) \rightarrow Cu(s) + ZnSO_4(aq)$$

   **(A)** Metallic zinc is the reducing agent.     **(B)** Metallic zinc is reduced.

   **(C)** Copper ion is oxidized.     **(D)** Sulfate ion is the oxidizing agent.

8. In this reaction, which substance behaves as the oxidizing agent?

   $$Pb(s) + PbO_2(s) + 2H_2SO_4(aq) \rightarrow 2PbSO_4(s) + 2H_2O(l)$$

   **(A)** Pb     **(B)** $PbSO_4$     **(C)** $PbO_2$     **(D)** $H_2SO_4$

9. Under certain conditions, $H_2O_2$ can act as an oxidizing agent; under other conditions, as a reducing agent. What is the best theoretical explanation for this?

   **(A)** $H_2O_2$ is a good bleaching agent.

   **(B)** Peroxides are stronger oxidizing agents than are oxides.

   **(C)** $H_2O_2$ will decolorize $KMnO_4$ solutions in the presence of an acid and will turn black lead sulfide to a white compound.

   **(D)** An atom within a compound can sometimes attain a more stable electronic structure either by gaining or by losing electrons.

10. Balance this ionic equation for a redox reaction, using only whole number coefficients.

$$? \ MnO_4^-(aq) + ? \ Fe^{2+}(aq) + ? \ H_3O^+(aq) \rightarrow ? \ Mn^{2+}(aq) + ? \ Fe^{3+}(aq) + ? \ H_2O(l)$$

What is the coefficient for $Fe^{2+}$ in the balanced equation?

(A)    1          (B)    3          (C)    4          (D)    5

11. When this oxidation-reduction equation is balanced in acidic solution, using only whole number coefficients, what is the coefficient for $S(s)$?

$$? \ Cr_2O_7^{2-}(aq) + ? \ H_2S(aq) \rightarrow ? \ Cr^{3+}(aq) + ? \ S(s)$$

(A)    4          (B)    3          (C)    2          (D)    1

12. Which of these ions is the best reducing agent?

| Standard Reduction Potentials | $E°$ |
|---|---|
| $Fe^{3+}(aq) + e^- \rightarrow Fe^{2+}(aq)$ | +0.77 V |
| $Cu^{2+}(aq) + e^- \rightarrow Cu^+(aq)$ | +0.15 V |

(A)    $Fe^{3+}$          (B)    $Fe^{2+}$          (C)    $Cu^{2+}$          (D)    $Cu^+$

13. A quantity of a powdered mixture of zinc and iron is added to a solution containing $Fe^{2+}$ and $Zn^{2+}$ ions, each at unit activity. What reaction will occur?

| Standard Reduction Potentials | $E°$ |
|---|---|
| $Fe^{3+}(aq) + e^- \rightarrow Fe^{2+}(aq)$ | +0.77 V |
| $Fe^{2+}(aq) + 2e^- \rightarrow Fe(s)$ | −0.44 V |
| $Zn^{2+}(aq) + 2e^- \rightarrow Zn(s)$ | −0.76 V |

(A)    Zinc ions will oxidize Fe to $Fe^{2+}$.          (B)    $Fe^{2+}$ ions will be oxidized to $Fe^{3+}$ ions.

(C)    Zinc ions will be reduced to zinc metal.          (D)    Zinc metal will reduce $Fe^{2+}$ ions.

14. Two metals are represented by the symbols **L** and **Z**. Which is the best oxidizing agent listed?

| Standard Reduction Potentials | $E°$ |
|---|---|
| $L^{2+}(aq) + 2e^- \rightarrow L(s)$ | +0.76 V |
| $Z^+(aq) + e^- \rightarrow Z(s)$ | +0.80 V |

(A)    **L**          (B)    $L^{2+}$          (C)    **Z**          (D)    $Z^+$

15. Which combination of reactants will produce the greatest voltage based on these standard electrode potentials?

| Standard Reduction Potentials | $E°$ |
|---|---|
| $Cu^+(aq) + e^- \rightarrow Cu(s)$ | +0.52 V |
| $Sn^{4+}(aq) + 2e^- \rightarrow Sn^{2+}(aq)$ | +0.15 V |
| $Cr^{3+}(aq) + e^- \rightarrow Cr^{2+}(aq)$ | −0.41 V |

(A)    $Cu^+$ and $Sn^{2+}$     (B)    $Cu^+$ and $Cr^{2+}$     (C)    Cu and $Sn^{2+}$     (D)    $Sn^{4+}$ and $Cr^{2+}$

16. What is the standard electrode potential for a voltaic cell constructed in the appropriate way from these two half-cells?

| Standard Reduction Potentials | $E°$ |
|---|---|
| $Cr^{3+}(aq) + 3e^- \rightarrow Cr(s)$ | −0.74 V |
| $Co^{2+}(aq) + 2e^- \rightarrow Co(s)$ | −0.28 V |

(A)    −1.02 V          (B)    0.46 V          (C)    0.64 V          (D)    1.02 V

17. What is the standard cell potential, $E°$, for this reaction?

$$3Mn(s) + 2AuCl_4^-(aq) \rightarrow$$
$$3Mn^{2+}(aq) + 2Au(s) + 8Cl^-(aq)$$

| Standard Reduction Potentials | $E°$ |
|---|---|
| $Mn^{2+}(aq) + 2e^- \rightarrow Mn(s)$ | −1.18 V |
| $AuCl_4^-(aq) + 3e^- \rightarrow Au(s) + 4Cl^-(aq)$ | +1.00 V |

(A)    −0.18 V          (B)    −2.18 V          (C)    +2.18 V          (D)    +5.54 V

18. What is the standard cell potential, $E°$, for this half-reaction?

    $$Pd^{2+}(aq) + 2e^- \rightarrow Pd(s)$$

| Standard Reaction Potentials | $E°$ |
|---|---|
| $Cu(s) + Pd^{2+}(aq) \rightarrow Cu^{2+}(aq) + Pd(s)$ | +0.650 V |
| $Cu^{2+}(aq) + 2e^- \rightarrow Cu(s)$ | +0.337 V |

    (A)    +0.987 V     (B)    –0.987 V     (C)    +0.313 V     (D)    –0.313 V

19. The standard cell potential, $E°$, for this reaction is 0.79 V.

    $$6I^-(aq) + Cr_2O_7^{2-}(aq) + 14H^+(aq) \rightarrow 3I_2(aq) + 2Cr^{3+}(aq) + 7H_2O(l)$$

    What is the standard potential for $I_2(aq)$ being reduced to $I^-(aq)$ given that the standard reduction potential for $Cr_2O_7^{2-}(aq)$ changing to $Cr^{3+}(aq)$ is +1.33 V?

    (A)    +0.54 V     (B)    –0.54 V     (C)    +0.18 V     (D)    –0.18 V

20. During the electrolysis of an aqueous solution of $CuSO_4$ using inert electrodes,

    (A) the anode loses mass and the cathode gains mass.

    (B) the mass of the anode decreases but the mass of the cathode remains constant.

    (C) the mass of the anode remains the same but the cathode gains mass.

    (D) the anode and the cathode neither gain nor lose mass.

21. Fluoride ion in aqueous solution are difficult to oxidize at the anode of an electrolytic cell because

    (A) the aqueous solutions of fluorides are nonconducting.

    (B) it is impossible to find the proper material from which to build the electrodes.

    (C) the fluorides are not very soluble.

    (D) oxygen is released from water in preference to fluorine.

22. Which products are formed during the electrolysis of a concentrated aqueous solution of sodium chloride?

    I. $Cl_2(g)$        II. $NaOH(aq)$        III. $H_2(g)$

    (A) I only                    (B) I and II only

    (C) I and III only        (D) I, II, and III

23. What mass of platinum could be plated on an electrode from the electrolysis of a $Pt(NO_3)_2$ solution with a current of 0.500 A for 55.0 s?

    (A)    27.8 mg     (B)    45.5 mg     (C)    53.6 mg     (D)    91.0 mg

24. A current of 5.00 A is passed through an aqueous solution of chromium(III) nitrate for 30.0 min. How many grams of chromium metal will be deposited at the cathode?

    (A)    0.027 g     (B)    1.62 g     (C)    4.85 g     (D)    6.33 g

25. A vanadium electrode is oxidized electrically. If the mass of the electrode decreases by 114 mg during the passage of 650 coulombs, what is the oxidation state of the vanadium product?

    (A)    +1     (B)    +2     (C)    +3     (D)    +4

**26.** In which case would the *least* number of faradays of electricity be required for the liberation of 1.0 g of free metal?

| Atomic Molar Masses | |
|---|---|
| K | 39.1 g·mol⁻¹ |
| Na | 23.0 g·mol⁻¹ |
| Cu | 63.5 g·mol⁻¹ |
| Ag | 107.9 g·mol⁻¹ |

    **(A)**      K from molten KOH          **(B)**      Na from molten NaCl

    **(C)**      Cu from aqueous $CuSO_4$        **(D)**      Ag from aqueous $AgNO_3$

**27.** How many minutes will be required to deposit 1.00 g of chromium metal from an aqueous $CrO_4^{2-}$ solution using a current of 6.00 amperes?

    **(A)**     186 min      **(B)**     30.9 min      **(C)**     15.4 min      **(D)**     5.15 min

**28.** Consider this reaction.

$$Sn^{2+}(aq) + 2Fe^{3+}(aq) \rightarrow Sn^{4+}(aq) + 2Fe^{2+}(aq) \qquad E° = 0.617 \text{ V}$$

| The Nernst Equation |
|---|
| $E = E° - \dfrac{0.0592}{n}\log(Q)$ |

What is the value of $E$ when $[Sn^{2+}]$ and $[Fe^{3+}]$ are equal to 0.50 M and $[Sn^{4+}]$ and $[Fe^{2+}]$ are equal to 0.10 M?

    **(A)**     0.069V      **(B)**     0.679 V      **(C)**     0.658 V      **(D)**     0.576 V

**29.** Consider this reaction.

$$4e^- + 4H^+(aq) + O_2(g) \rightleftharpoons 2H_2O(l) \qquad E° = 1.23 \text{ V}$$

| The Nernst Equation |
|---|
| $E = E° - \dfrac{0.0592}{n}\log(Q)$ |

Which statement is true if the hydrogen ion concentration is initially at 1.0 M and the initial pressure of oxygen gas is 1.0 atmosphere?

    **(A)**     Addition of a base should result in a value of $E$ which is less than 1.23 V.

    **(B)**     $n = 1$, since one mole of oxygen is being considered.

    **(C)**     $E$ is independent of the pH of the solution.

    **(D)**     $Q = \dfrac{[H_2O]^2}{[O_2][H^+]}$

**30.** Which of these changes will produce the most positive voltage for this half reaction in the direction written?

$$Co^{2+}(aq) + 2e^- \rightarrow Co(s) \qquad E° = -0.28 \text{ V}$$

    **(A)**     increasing the amount of solid Co

    **(B)**     decreasing the amount of solid Co

    **(C)**     increasing the concentration of $Co^{2+}$

    **(D)**     decreasing the concentration of $Co^{2+}$

## *Answers to Study Questions*

| | | | | | |
|---|---|---|---|---|---|
| 1. | A | 5. | B | 9. | B |
| 2. | C | 6. | B | 10. | B |
| 3. | C | 7. | C | | |
| 4. | B | 8. | C | | |

## *Answers to Practice Questions*

| | | | | | |
|---|---|---|---|---|---|
| 1. | A | 11. | B | 21. | D |
| 2. | D | 12. | D | 22. | D |
| 3. | B | 13. | D | 23. | A |
| 4. | C | 14. | D | 24. | B |
| 5. | C | 15. | B | 25. | C |
| 6. | A | 16. | B | 26. | D |
| 7. | A | 17. | C | 27. | B |
| 8. | C | 18. | A | 28. | B |
| 9. | D | 19. | A | 29. | A |
| 10. | D | 20. | C | 30. | C |

# Descriptive Chemistry / Periodicity

This section of the general chemistry course deals with a variety of physical and chemical properties. Developing a general sense of color, odor, solubility, melting and boiling points for common chemical substances will be addressed throughout the course, but may be emphasized together with the study of periodic properties. A trend that is emerging in general chemistry exams is the inclusion of more descriptive chemistry, particularly if there are applications to consumer chemistry, environmental chemistry, or industrial chemistry.

Understanding the wealth of information found in the organization of the periodic table is a central skill for general chemistry. You will always have a periodic table available for ACS exams, and likely for most classroom tests as well. Therefore, knowing the trends within the periodic table will allow prediction of properties, even for unfamiliar elements. Relative sizes of atoms and ions, trends in ionization energy, and trends in electronegativity are all important to understanding the behavior of elements. The differences between metals and nonmetals and their reactions are also based on periodic trends. Trends within families and trends within periods can both reveal much about the physical properties and chemical reactions expected for the elements.

As you work through this section, please use the periodic table found in the Appendix of this study guide for reference.

## Study Questions

**DP-1.** What property of aluminum makes it particularly useful as a lining for steel tank cars transporting nitric acid?

    **(A)** It forms an amphoteric hydroxide.

    **(B)** It is a metal with a very low density.

    **(C)** It is below hydrogen in the activity series.

    **(D)** It forms an adherent and insoluble oxide coating.

*Knowledge Required:* (1) The properties of aluminum metal. (2) Understanding how properties are related to particular uses.

*Thinking it Through:* Aluminum is a metal familiar to all, but you may not have considered this particular use for aluminum. There will be a chemical reaction if steel, which is mainly iron with varying amounts of carbon and other elements, were to come in contact with nitric acid. In considering each statement about aluminum, it is first necessary to decide if each statement is *true*. If so, the next question to consider is whether or not that statement gives a *reason* that aluminum can effectively keep the steel and nitric acid from coming into contact and therefore reacting. Although choice **(A)** is a true statement based on aluminum's position in Group IIIA on the periodic table, it does not offer a reasonable *explanation* for this industrial use. Aluminum's amphoteric hydroxide, one that is capable of acting as an acid or a base, must not be present for it would act as a base and react with the nitric acid. Choice **(B)** also is a true statement. Aluminum does have a high strength-to-mass ratio, and that is helpful in this and other uses but this property is not explanation of the primary reason that aluminum is used to line steel tank cars. Choice **(C)** is not a true statement. Aluminum is above hydrogen in the activity series, but does not often act as such an active metal for the reason given in correct choice **(D)**. Aluminum finds wide use in consumer and industrial applications because it has the unusual ability to protect itself with a tough and adherent oxide coating.

**DP-2.** Which gas is most soluble in water?

    **(A)** ammonia, $NH_3$    **(B)** hydrogen, $H_2$    **(C)** methane, $CH_4$    **(D)** nitrogen, $N_2$

*Knowledge Required:* (1) The solubility of common gases. (2) Prediction of solubility based on the principle that "like dissolves like".

***Thinking it Through:*** This question can be answered quickly based on the recall of the solubility properties of the common gases. Choice **(A)** is correct. You may recall seeing the "ammonia fountain" demonstrated or perhaps you have tried it yourself in the laboratory. This popular chemical demonstration depends on the high solubility of ammonia in water. Another approach is to look at the underlying cause of solubility. The solubility of a gas in water is based on the extent of the interactions of the gas molecules with highly polar water molecules. Of the gases given in this question, only ammonia is a polar molecule, and therefore only ammonia exhibits the "like dissolves like" interactions that bring it into solution. The other three choices can be eliminated because $H_2$, $CH_4$, and $N_2$ are all nonpolar molecules, limiting their solubility in water. Any solubility is due to the weak, temporary dipoles that can form even in nonpolar molecules.

**Note:** Other questions dealing with descriptive chemistry may ask about other important properties of gases, such as their ability to burn, support combustion, or take part in atmospheric reactions.

---

**DP-3.**    Which is a characteristic property of transition elements?

(A)    colorless ions                    (B)    variable oxidation number

(C)    formation of strong bases         (D)    non-metallic behavior

---

***Knowledge Required:*** (1) The meaning of the term *transition elements*. (2) The general properties of transition elements, based on their position on the periodic table.

***Thinking it Through:*** Between the main groups IIA and IIIA on the periodic table, there are ten vertical columns containing the transition elements. The transition elements are termed the B subgroups in most texts. The 1-18 designations recommended by the International Union of Pure and Applied Chemistry (IUPAC) are also given for your consideration, although this numbering system is not yet in widespread use in most texts and exams. Note that the transition elements are found in groups 3-12 in this system.

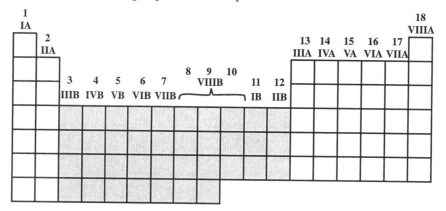

The transition elements have electronic structures in which the *d* subshell is only partially filled in any common oxidation state. This group of elements has a number of characteristic properties that set them apart from the main-group elements. Most of the compounds of the transition elements are colored, and many are *paramagnetic*, meaning there are unpaired *d* electrons. Both of these are the result of the partially filled *d* subshell. Most of the B group elements have variable oxidation states, although there are exceptions particularly in the IIB and IIIB groups. In the main group elements, only the heavier elements display a variety of oxidation states in common compounds. All of the transition elements are metals; the B groups are sometimes called the *transition metals*. They are all hard solids with relatively high melting and boiling points, compared with the main group elements.

Given the choices in this question, **(A)** is incorrect because most ions of transitions are colored. Choice **(B)** is correct, for the variation in oxidation state is a common characteristic of most transitions. Choice **(C)** is a property of the very active metals of the IA and IIA groups, but not of the transition metals. Choice **(D)** is incorrect because the transition elements all exhibit metallic rather than nonmetallic characteristics.

**Note:** Strictly defined the term transition elements may be *d*-block or *f*-block elements. However, it is common to refer to the *d*-block elements as the transition elements and to use the term inner transition elements for the *f*-block transition elements.

**DP-4.** Which statement best describes why soap forms a scummy precipitate in hard water?

    **(A)** Soap cannot hydrolyze in hard water, so it precipitates.

    **(B)** Soap forms a precipitate with the metallic ions in the water.

    **(C)** Soap requires an acidic solution to dissolve, but hard water is alkaline.

    **(D)** Soap removes the dirt in the hard water in the form of a precipitate.

*Knowledge Required:* (1) The meaning of the term *hard water*. (2) The chemical behavior of soap in hard water.

*Thinking it Through:* Hard water contains dissolved ions of $Ca^{2+}$ and $Mg^{2+}$. In some parts of the country, there may be dissolved iron as well. There are several potential problems if the mineral content of water is too high, including the interaction with soap to form a undesirable curd-like precipitate that settles on clothing, dishes, or you. Soap is the sodium salt of the long-chain organic acid, stearic acid. Perhaps you have even made soap from lye, a source of sodium hydroxide, and animal fat, a common source of stearic acid. This is the structure of soap.

$$CH_3(CH_2)_{16}\overset{\displaystyle O}{\overset{\|}{C}}-O^-\ Na^+$$

The cleaning ability of soap can be explained by the presence of both a charged region and a nonpolar hydrocarbon region on the stearate ion. The charged region of the ion is soluble in water and the nonpolar hydrocarbon rest of the ion is soluble in grease and oils. These are both examples of "like dissolves like" and both are taking place on the *same* ion. Therefore the charged region drags the nonpolar end of the molecule together with grease and oils into solution, at least for long enough to be carried away with the rinse water. However, if the stearate ion finds $Ca^{2+}$ and $Mg^{2+}$ ions in the water, the ions will precipitate, forming calcium stearate and magnesium stearate. The precipitate often is quite slimy because residual grease and oil are still attached to the hydrocarbon end of the stearate ion, even when it precipitates.

$$2CH_3(CH_2)_{16}\overset{\displaystyle O}{\overset{\|}{C}}-O^-\ +\ Ca^{2+}\ \longrightarrow\ Ca\left(CH_3(CH_2)_{16}\overset{\displaystyle O}{\overset{\|}{C}}-O\right)_{2\ (s)}$$

Correct choice **(B)** describes this process. Choice **(D)** is an incorrect variation. It is not the "dirt" in the hard water that is precipitating with the soap, it is previously *dissolved ions*. Choice **(C)** gives incorrect assumptions about the both the acidic nature of solutions required by soap and the basic property of hard water. Choice **(A)** uses a somewhat familiar term, hydrolysis, but it is used incorrectly in this case. Hydrolysis for the stearate ion describes the interaction of the ion with water to produce the conjugate acid and release hydroxide ions, not a factor in forming the observed precipitate.

**DP-5.** Which is a result of the process of photosynthesis?

    **(A)** Nitrogen is changed to ammonia.

    **(B)** Carbohydrates and oxygen gas are formed.

    **(C)** The carbon in decaying organic matter is released as carbon dioxide.

    **(D)** Energy from sunlight is converted to heat given off by growing green plants.

*Knowledge Required:* (1) The meaning of the term *photosynthesis*. (2) Connecting meaning to the chemical equation representing photosynthesis.

*Thinking it Through:* There are many processes from organic and biochemistry that *may* be discussed in a general chemistry class, but the most likely is the process of photosynthesis. You have likely already encountered the process in your biology courses, but in studying chemistry, the equation is often balanced for practice and the energy changes discussed in the energetics unit of study. It also is likely to be presented as an example when discussing changes in entropy or during consideration of environmental issues because the overall reaction removes the major greenhouse gas, carbon dioxide, from the atmosphere. Although it is actually a far more complex process involving many steps, this is the simplest representation of the overall process of photosynthesis.

$$6CO_2(g) \quad + \quad 6H_2O(l) \quad \xrightarrow{\text{chlorophyll, energy from the sun}} \quad C_6H_{12}O_6(aq) \quad + \quad 6O_2(g)$$

carbon dioxide gas + water → glucose + oxygen gas

The result of the process of photosynthesis is to change carbon dioxide and water into glucose, which is a simple carbohydrate, and oxygen gas. This is correct choice (**B**). Choice (**A**) can be easily dismissed, as nitrogen is not directly involved in photosynthesis. You might be incorrectly remembering some other processes involving nitrogen from a previous biology course if you chose (**A**). Choice (**C**) describes a *reverse* process from photosynthesis, aerobic decay of organic matter. Choice (**D**) may attract your attention because energy from the sun is used to power this reaction. It is incorrect overall, however, because this process is not exothermic but rather *endothermic*. The sunlight's energy is continually needed to fuel the photosynthetic process. Photosynthesis cannot take place in the absence of this energy. Another proof of this is that the reverse process described in choice (**C**) is exothermic. You may know from experience that considerable heat is released as the organic material in your compost pile undergoes decomposition.

---

**DP-6.** If the formula of an oxide of element **X** is $X_2O_3$, what is the formula of the chloride of **X**?

(**A**) $XCl_3$     (**B**) $XCl$     (**C**) $X_3Cl$     (**D**) $XCl_6$

*Knowledge Required:* (1) Using the periodic table to predict a chemical formula. (2) Review of Mendeleev's original organization of the periodic table.

*Thinking it Through:* Although the arrangement of elements in the modern periodic table can be explained by the periodic nature of ground-state electron configurations, Mendeleev had no such knowledge in 1869 when he proposed his organization of the elements. His table was based on increasing atomic weights and the elements were grouped by their ability to form oxides with similar formulas. Even though we now organize the table based on increasing atomic number, there have been very few changes in Mendeleev's original organization except to include new elements as they have been discovered. It is this proven predictive power of the periodic table that is used in this problem. If the unknown element X forms a compound with the formula $X_2O_3$, the oxidation state of **X** must be +3 so that 2 atoms will equal the total negative charge of 3(−2) contributed by the oxygen. Therefore when the more metallic **X** is combined with chlorine in a binary compound, it will take 3 chlorines, each with a −1 oxidation number, to equal the +3 of element **X**. The correct choice is (**A**). The incorrect choice (**B**) does not appear to have been based on any real reason for the 1:1 ratio of **X** to Cl. Incorrect choice (**C**) has misinterpreted the meaning of the +3 oxidation state, and incorrect choice (**D**) may have been picked if chloride, Cl⁻, was confused with chlorine, $Cl_2$.

**Note:** The only thing we can definitely conclude about X is that it has an oxidation state of +3. Element **X** is not *necessarily* in group IIIA, although it does not hurt *in this particular problem* to make that assumption. That could mean X was a transition metal or even a nonmetal with that oxidation state.

---

**DP-7.** When the three elements S, Se, and Cl are arranged in order of increasing atomic radius, which is the correct order?

(**A**) Se < S < Cl      (**B**) S < Cl < Se

(**C**) Cl < S < Se      (**D**) S < Se < Cl

*Knowledge Required:* (1) The periodic trends in atomic radii within groups and periods.

*Thinking it Through:* Even though electron density extends far beyond the nucleus of an atom, the usual limit for measuring the size of the electron cloud includes about 90% of the total electron density. Then, rather than giving the volume, atomic size is often expressed by giving its atomic radius, which is *proportional* to the volume. There is a regular trend in the atomic radii of the main group elements based on their positions in the periodic table. This table expresses the expected trends. Your textbook may include a plot showing how atomic radius varies as a function of atomic number, another good way to consider the periodic nature of the atomic radius.

Looking first at the trend within any vertical column or group, it is quite easy to understand that if an element has electrons at successively greater energy levels, the atomic radius of an element will be larger. Selenium, being found in the 4th period of Group VIA, has a larger atomic radius than S, found in the 3rd period of Group VIA. This eliminates choice (A), for Se is indicated as being smaller than S, which cannot be correct.

Looking at the trend within any horizontal row or period, it is perhaps *not* as intuitive to understand the trend. The atomic radius increases *from right to left* through the main groups of the same period. This may seem counterintuitive because the elements are getting heavier going left to right. Balancing that line of reasoning is the fact that electrons are being added at the *same* energy level, not successively higher energy levels as was the case with the trend within groups. Within a period, the most important determinant of atomic radius will be how effective the positively charged nucleus can attract electrons. The effective nuclear charge increases from left to right within a period, as the number of protons increases steadily. Electrons being added to the *same* subshell are not very effective at shielding the pull of the positively charged nucleus. Therefore, the electrons can be pulled more tightly to the nucleus in chlorine (17 protons, electronic distribution [Ne] $3s^2 3p^6$) than will be the case for sulfur (16 protons, electronic distribution [Ne] $3s^2 3p^5$). This makes the atomic radius of chlorine *smaller* than the atomic radius for sulfur. Choices (B) and (D) are now eliminated for they both show the opposite trend. The correct response is (C) with chlorine smaller than sulfur, which is smaller than selenium.

---

**DP-8.** In which pair of species is the first member *larger* than the second member?

(A) $Li^+$ and $Be^{2+}$    (B) $Li^+$ and $Na^+$    (C) $Li^+$ and Li    (D) Be and Mg

---

*Knowledge Required:* (1) The periodic trends in atomic radii. (2) The periodic trends in ionic size. (3) The relationship of the size of an ion to its atom.

*Thinking it Through:* This question asks about several different size trends. Atomic size trends were addressed in study problem **DP-7**, so choice (D) can be discarded because Be (group IIA, period 2) is *smaller* than Mg (group IIA, period 3). Choice (C) requires consideration of how the size of a metal ion compares with the corresponding atom. Lithium atom has 3 protons and 3 electrons but lithium ion has 3 protons and only 2 electrons. This means the constant nuclear charge of +3 will be more successful in pulling electrons close to the nucleus in the lithium ion, making it smaller. All metal ions will be *smaller* than their corresponding atoms for similar reasons, eliminating choice (C) from further consideration. Choices (A) and (B) require comparison of ion sizes. Again, it is useful to compare the relative number of protons and electrons available.

$Li^+$    3 protons      2 electrons

$Be^{2+}$    4 protons      2 electrons

If 4 protons are pulling on 2 electrons, the ion will be smaller than if only 3 protons are attracting the same number of electrons. This makes choice (**A**) the correct answer. Species with the same number of electrons are termed *isoelectronic*, and this may be the term used in other questions. In choice (**B**), this is the comparison of the relative number of protons and electrons available.

$Na^+$     11 protons      10 electrons

$Li^+$     3 protons      2 electrons

Lithium and sodium atoms are both in group IA, making the sodium atom larger than the lithium atom. This same trend is repeated in the ions, for each has lost just one electron in forming the ion. Choice (**B**) is also incorrect.

**Note:** Although not used in this question, all ions of nonmetals will be *larger* than their corresponding nonmetallic atom. The constant nuclear charge will be pulling on a greater number of electrons in the negatively charged ion, always making the ion larger than the corresponding atom.

---

**DP-9.**    Which of these elements is predicted to be the most active nonmetal?

| Unknown Elements | Electronegativity Values |
| --- | --- |
| Q | 0.9 |
| R | 1.0 |
| T | 2.8 |
| X | 3.0 |

(A)    Q       (B)    R       (C)    T       (D)    X

*Knowledge Required:* (1) The definition of electronegativity. (2) The relationship of electronegativity values to chemical behavior.

*Thinking it Through:* Electronegativity gives an indication of the relative ability of an atom to attract the electrons in a chemical bond. Being a relative value, it has no units and the values range between 0 and 4. Metals typically have rather low values of electronegativity because they tend to *lose* electrons in a chemical reaction, not attract additional electrons. Nonmetals tend to have higher values, for they are more effective in attracting electrons in a chemical bond. The element with the *highest* electronegativity value of 4.0 is fluorine in period 2 of group VIIA. The lowest values are expected in period 7 of group IA. Electronegativity is not defined for the noble gases, group VIIIA; they do not commonly form bonds and so the relative ability to attract bonded electrons is not given. The general trends are shown with this periodic table. A few key values have been indicated to emphasize the trends.

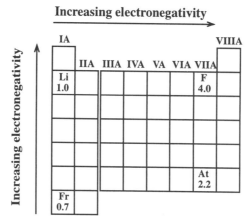

Of these unknown elements, **X** has the highest electronegativity (3.0), making it most likely to attract electrons and behave chemically as a nonmetal. This is choice (**D**). Both nitrogen (group VA, period 2) and chlorine (group VIIA, period 3) have an electronegativity of 3.0, and both are nonmetals. Element **Q** with the lowest electronegativity of the group, is definitely a metal and probably in group IA. While it is not necessary to identify the particular element, the electronegativity matches that of sodium (group IA, period 3) with a value of 0.9. The other two unknowns, being intermediate values, should not be considered as possible answers.

**Note:** The trends in electronegativity values run in the opposite directions from the trends observed for atomic radii; values are undefined for the noble gases.

**DP-10.**   Arrange the elements Li, Ne, Na, and Ar in increasing order of the energy required to remove the first electron from their respective gaseous atoms.

**(A)**   Na < Li < Ar < Ne

**(B)**   Li < Na < Ar < Ne

**(C)**   Na < Li < Ne < Ar

**(D)**   Ar < Ne < Na < Li

*Knowledge Required:* (1) The definition of *ionization energy.* (2) The periodic trends in first ionization energies within groups and periods.

*Thinking it Through:* Ionization energy is the amount of energy required to remove an electron from a gaseous atom or ion. This process *always* takes energy, for it separates a negatively charged electron from the positively charged nucleus. The first electron to be removed will be the one requiring the *least* expenditure of energy, and will be the one furthest from the nucleus of the atom. It always takes more energy to remove an electron from a smaller atom, so the trends in first ionization energies within a group or period are exactly the *reverse* of the trends previously observed for increasing atomic size. Within a group, the size of the atom increases from top to bottom, so first ionization energy increases from bottom to top. Within a period, size of the atom increases from right to left. The first ionization energy increases from left to right and closely follows the change from metals to nonmetals. Remember that nonmetals tend to *gain* electrons in forming a stable chemical bond, so the amount of energy required to separate the electron from the nucleus rises sharply across a given period. The maximum first ionization energy value will be for the group VIII noble gas in each period. These atoms require a very large amount of energy to remove an electron from the stable octet of outer electrons. These trends are summarized in this table, which includes the atoms under consideration in this question.

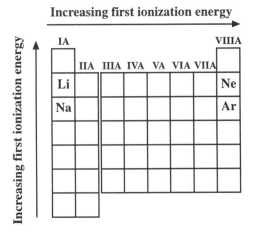

Of the four gaseous atoms under consideration in this problem, neon will require the most energy to remove an electron. It is both to the right of lithium in period 2 and above argon in group VIIIA. This observation alone eliminates choices **(C)** and **(D)** from further consideration. In choice **(C)**, argon is identified as the atom with the largest first ionization energy, indicating inaccurate knowledge of the trend within a group. In choice **(D)**, lithium is listed as the atom with the largest first ionization energy, indicating inaccurate knowledge of the trend within a period. To decide between **(A)** and **(B)**, remember that lithium is smaller than sodium and that the trend within a group is for increasing ionization energy being associated with smaller atoms. The same is true for argon and neon, making the correct choice **(A)**. Choice **(B)** has reversed the trend within the alkali metals, but has the noble gas atoms correctly ordered.

**Note:** Trends in *successive* ionization energies are also related to the size of the atom or ion under consideration. It will take considerably more energy, for example to remove the second electron from a sodium atom than it did for the first, because the sodium ion is considerably smaller than the sodium atom.

## Practice Questions

1. Which metal reacts with concentrated NaOH to produce hydrogen gas?

   (A)   aluminum   (B)   copper   (C)   iron   (D)   magnesium

2. Which is the most satisfactory solvent for alumina, $Al_2O_3$, in the electrolytic process for the preparation of aluminum?

   (A)   liquid water, $H_2O$   (B)   molten cryolite, $Na_3AlF_6$

   (C)   molten bauxite, $Al_2O_3$ with other oxides   (D)   molten sodium chloride, NaCl

3. Which of these gases can be most efficiently collected by the displacement of water?

   (A)   ammonia   (B)   carbon dioxide   (C)   oxygen   (D)   hydrogen chloride

4. Which gas is *least* soluble in water?

   (A)   $H_2$   (B)   $CO_2$   (C)   $NH_3$   (D)   $SO_2$

5. Which gas is the predominant nitrogen-containing pollutant in the atmosphere?

   (A)   $N_2$   (B)   $NH_3$   (C)   $NO_2$   (D)   $N_2O_5$

6. Helium is preferable to hydrogen for filling balloons because helium is

   (A)   less dense.   (B)   less expensive.

   (C)   easier to obtain.   (D)   chemically inert.

7. Transition metals typically have all of these characteristics *except*

   (A)   forming colored compounds.

   (B)   showing a variety of oxidation states.

   (C)   having individual atoms with one or more unpaired electrons.

   (D)   having low melting points in the elemental state.

8. Which compound is *not* expected to be colored?

   (A)   $CuCl_2$   (B)   $K_2Cr_2O_7$   (C)   $TiO_2$   (D)   $NiSO_4$

9. The members of which pair are most similar in color?

   (A)   chlorine, $Cl_2(g)$, and bromine, $Br_2(l)$

   (B)   silver chloride, AgCl(s), and silver sulfide, $Ag_2S(s)$

   (C)   copper(II) chloride, $CuCl_2(s)$, and calcium chloride, $CaCl_2(s)$

   (D)   potassium permanganate, $KMnO_4(aq)$, and iodine, $I_2(g)$

10. Which natural process tends to change the free nitrogen of the air into "fixed" nitrogen?

   (A)   lightning storms   (B)   photosynthesis

   (C)   respiration of animals   (D)   rusting of iron

11. When glucose, $C_6H_{12}O_6$, is completely oxidized with excess oxygen, what are the products?

    (A)  $H_2O_2$ and $CO_2$          (B)  $H_2O$ and $CO_2$

    (C)  $H_2O_2$ and CO           (D)  $H_2O$ and CO

12. Which ion may form a scummy precipitate with ordinary soap?

    (A)  $HCO_3^-$     (B)  $CO_3^{2-}$     (C)  $Na^+$     (D)  $Ca^{2+}$

13. Which statement best reflects the organization of the modern periodic table?

    (A)  Elements are always arranged in order of increasing atomic weights.

    (B)  Metallic properties increase going from bottom to top in a family of elements.

    (C)  Nonmetallic properties tend to predominate for elements at the far right portion of the table.

    (D)  Each transition element is placed in the column of the main group element that it most closely resembles.

14. If gallium, atomic number 31, combines with selenium, atomic number 34, what is the most likely formula based on your knowledge of the periodic nature of the elements?

    (A)  GaSe     (B)  $GaSe_2$     (C)  $Ga_2Se$     (D)  $Ga_2Se_3$

15. An element forms a basic oxide with the formula XO and a hydride with the formula $XH_2$. The hydride reacts with water to give hydrogen. The element X could be

    (A)  K     (B)  Ca     (C)  N     (D)  O

16. The elements **X**, **Y**, and **Z** form these compounds: $XCl_4$, $XZ_2$, and YO. What formula would you predict for the compound formed between **Y** and **Z**? Assume normal oxidation states for chlorine and oxygen.

    (A)  **YZ**     (B)  $YZ_2$     (C)  $Y_2Z$     (D)  $YZ_3$

17. The size of metal atoms

    (A)  generally increases progressively from top to bottom in a group in the periodic table.

    (B)  generally increases progressively from left to right in a period.

    (C)  are smaller than those of the corresponding ions.

    (D)  do not change upon losing electrons.

18. Which of these elements has the *smallest* atomic radius?

    (A)  fluorine     (B)  chlorine     (C)  bromine     (D)  iodine

19. Which of these atoms will be the *smallest*?

    (A)  Si $(Z = 14)$     (B)  P $(Z = 15)$     (C)  Ge $(Z = 32)$     (D)  As $(Z = 33)$

20. The radii of the ions in this series decrease because

| Ion | Ionic Radii |
|---|---|
| $Na^+$ | 0.095 nm |
| $Mg^{2+}$ | 0.065 nm |
| $Al^{3+}$ | 0.050 nm |

   (A)   the elements are in the same period.

   (B)   the effective nuclear charge is increasing.

   (C)   the atomic radius of Na decreases from Na to Al.

   (D)   the first ionization energies increase from Na to Al.

21. The species $F^-$, Ne, and $Na^+$ all have the same number of electrons. Which is the predicted order when they are arranged in order of decreasing size (largest first)?

   (A)   $F^- > Ne > Na^+$

   (B)   $Ne > Na^+ > F^-$

   (C)   $Na^+ > F^- > Ne$

   (D)   $F^- > Na^+ > Ne$

22. Which isoelectronic ion is the *smallest*?

   (A)   $Al^{3+}$      (B)   $Na^+$      (C)   $F^-$      (D)   $O^{2-}$

23. What happens when a bromine atom becomes a bromide ion?

   (A)   A positive ion is formed.

   (B)   The bromine nucleus acquires a negative charge.

   (C)   The bromide ion is larger than the bromine atom.

   (D)   The atomic number of bromine is decreased by one.

24. Which ion has the largest radius?

   (A)   $Cl^-$      (B)   $F^-$      (C)   $K^+$      (D)   $Cu^{2+}$

25. Which of these elements has the highest electronegativity?

   (A)   oxygen      (B)   iodine      (C)   cesium      (D)   lithium

26. Which element is more electronegative than arsenic and less electronegative than sulfur?

   (A)   chlorine      (B)   phosphorus      (C)   tin      (D)   oxygen

27. List the elements Ca, Si, and K in order of *increasing* electronegativity.

   (A)   $Ca < K < Si$

   (B)   $K < Si < Ca$

   (C)   $Si < Ca < K$

   (D)   $K < Ca < Si$

28. Which pair of elements is listed in order of decreasing first ionization energy?

   (A)   Na, Mg      (B)   Mg, Al      (C)   Al, Si      (D)   Si, P

29. When the species $F^-$, $Na^+$, and Ne are arranged in order of increasing energy for the removal of an electron, what is the correct order?

    (A)     $F^- < Na^+ < Ne$               (B)     $F^- < Ne < Na^+$

    (C)     $Na^+ < Ne < F^-$               (D)     $Ne < F^- < Na^+$

30. The first three ionization energies of an element **X** are 590, 1145, and 4912 kJ·mol$^{-1}$. What is the most likely formula for a stable ion of **X**?

    (A)     $X^+$          (B)     $X^{2+}$          (C)     $X^{3+}$          (D)     $X^-$

## Answers to Study Questions

| | | | | | |
|---|---|---|---|---|---|
| 1. | D | 5. | B | 9. | D |
| 2. | A | 6. | A | 10. | A |
| 3. | B | 7. | C | | |
| 4. | B | 8. | A | | |

## Answers to Practice Questions

| | | | | | |
|---|---|---|---|---|---|
| 1. | A | 11. | B | 21. | A |
| 2. | B | 12. | D | 22. | A |
| 3. | C | 13. | C | 23. | C |
| 4. | A | 14. | D | 24. | A |
| 5. | C | 15. | B | 25. | A |
| 6. | D | 16. | A | 26. | B |
| 7. | D | 17. | A | 27. | D |
| 8. | C | 18. | A | 28. | B |
| 9. | D | 19. | B | 29. | B |
| 10. | A | 20. | B | 30. | B |

# Laboratory Chemistry

The laboratory is a place where we focus on techniques, experiments, and verification of theories. Results from laboratory investigations help answer the question, *"How do we know?"* This section deals with instruments, techniques, measurements, significant figures, interpretation of data, and safety. Although your laboratory experiences may be quite different, the general topics addressed here should enable you to demonstrate your knowledge of chemistry as practiced by chemists in the laboratory.

## Study Questions

**LC-1.** The mass spectrometer may be used to

    **(A)** detect the presence of isotopes.

    **(B)** increase the mass of atomic nuclei.

    **(C)** electrolyze water.

    **(D)** separate different wavelengths of light.

*Required Knowledge:* (1) Nature of a mass spectrometric data. (2) Definition of isotopes.

*Thinking it Through:* Mass spectrometers separate particles based on their charge/mass ratios. Particles having identical charges, but differing in mass, produce separate peaks. Knowing that isotopes *differ in mass,* because of differing numbers of neutrons in their nuclei, provides the key to choosing **(A)**. Choice **(B)** could only be accomplished by altering the nucleus of an atom, and would require an instrument that can change nuclear structure, such as a particle accelerator. Choice **(C)** only requires a direct electric current, and choice **(D)** might be incorrectly chosen because of possible confusion with optical spectrographs that contain prisms or diffraction gratings to separate light into its component wavelengths.
*Note*: It is not necessary to have *used* a mass spectrometer to be successful with this question. Texts often contain a schematic diagram of this instrument, usually in the section where evidence for the existence of isotopes is first discussed.

**LC-2.** Which laboratory measurement is expressed to the appropriate number of significant figures?

    **(A)** 6.12 °C on a thermometer graduated in degrees

    **(B)** 35 s on a timer graduated in tenths of seconds

    **(C)** 0.782 cm on a ruler graduated in millimeters

    **(D)** 22.57 mL in a buret graduated in tenths of milliliters

*Required Knowledge:* (1) Limitations of interpolation between graduations in measuring instruments. (2) Correct use of significant figures in reporting measurements.

*Thinking it Through:* Measurements should always be reported with the number of significant figures appropriate to the measuring device. In general, it is possible to estimate and report one digit beyond marked graduations. Therefore, it is possible to read a buret that is graduated in tenths of milliliters to the nearest hundredth of a mL. Choices **(A)** and **(C)** *overstate* the precision of the measuring device; Choice **(B)** *understates* the precision that can be achieved with each measuring device.

**LC-3.** The mass of an object was determined four times using four different balances: 7.613 g, 7.618 g, 7.615 g, and 7.618 g. The accepted mass of the object was 7.433 g. Which statement best describes the results of the experiment?

**(A)** poor precision and poor accuracy

**(B)** good precision and good accuracy

**(C)** poor precision and good accuracy

**(D)** good precision and poor accuracy

*Required Knowledge:* (1) Qualitative comparison of experimental results to accepted values. (2) Meaning of the terms *precision* and *accuracy*.

*Thinking it Through:* *Precision* is a measure of how close experimental results are to each other. *Accuracy* is a measure of how closely either a single measurement or the average of a group of measurements agrees with the accepted or "true" value. Inspection of the given experimental results shows they are quite close to each other, differing only in the last digit. This observation of good precision eliminates choices **(A)** and **(B)**. However, the experimental values are not centered on the true value of 7.433 g, leading to the conclusion that the results do not appear to be very accurate. This makes choice **(D)** the correct response, rather than **(C)**. In other questions, you may be asked to make a *quantitative* comparison of the experimental values to the true value. This is called the percent error or the percent difference. This requires finding the average result and comparing the absolute value of the *difference* between the average result and the accepted value with the accepted value, multiplying by 100 to change the ratio to a percent.

$$\% \text{ error} = \left( \frac{|(\text{accepted value} - \text{experimental value})|}{\text{accepted value}} \right) \times 100$$

*Note:* If a much larger number of measurements were taken for the mass of an object, these experimental results *might be* part of a more accurate average value.

**LC-4.** The relationship between the volume of a liquid contained in a graduated cylinder as a function of the number of submersible glass spheres added to the cylinder is shown in the graph. What is the average volume of one sphere?

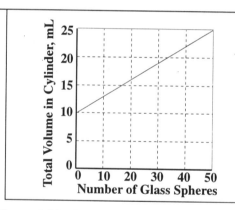

**(A)** 3.3 mL **(B)** 2.0 mL **(C)** 0.50 mL **(D)** 0.30 mL

*Required Knowledge:* (1) Interpretation of experimental data in graphical format. (2) Processing graphical data to draw a conclusion.

*Thinking it Through:* There is 10 mL of a liquid in the cylinder before any spheres are added. The volume of the liquid rises as spheres are added and the line shows a direct proportion between the number of spheres and the *change* in volume. After 50 spheres have been added, the volume has changed by 15 mL, from which you can calculate that the average volume of the sphere is 15 mL / 50 spheres = 0.30 mL, which is choice **(D)**. Both incorrect choices **(B)** and **(C)** have used the *total* volume of 25 mL, rather than the *change* in volume, 15 mL. Choice **(A)** can be eliminated based on its size. *Note:* Assuming the liquid is water, how would this graph change if sugar cubes were used rather than glass spheres?

**LC-5.** In an experiment to determine the molar mass of an unknown solid acid, the acid is titrated with a standard solution of NaOH. Which procedural error(s) will result in a calculated molar mass that is too low?

| Procedural Mistakes | |
|---|---|
| I | using a NaOH solution that has absorbed $CO_2$ |
| II | rinsing the buret with distilled $H_2O$ rather than NaOH solution before titrating |
| III | allowing some of the solid acid to stick to the side of the flask |

(A)    I only

(B)    III only

(C)    I and II only

(D)    II and III only

*Required Knowledge:* (1) The laboratory procedures for acid–base titrations. (2) The mathematical effect each possible mistake will have on the calculated result.

*Thinking it Through:* The starting points of this procedure are to weigh out a sample of the solid acid and to prepare a buret containing standardized NaOH solution. Titration continues until the end point is reached, using an appropriate indicator. The number of moles hydroxide ion used in the titration is equal to the number of moles of ionizable hydrogen ion in the acid. This is the mathematical expression, assuming the solid acid is monoprotic.

$$V_{NaOH} \times M_{NaOH} = \frac{g_{\text{solid acid}}}{M_{\text{solid acid}}}$$

Inspection of this equation shows that both volume and molarity of the base are *inversely* proportional to the molar mass, *M*. Conditions that would produce a calculated molar mass that is too low include having either the volume or the molarity of the sodium hydroxide solution too large. Factor **I** will mean a *larger volume* of NaOH must be used to reach the endpoint as some of the NaOH has reacted with the $CO_2$. Factor **II** will also mean a *larger volume* of NaOH must be used to reach the end point for the NaOH is not as concentrated as anticipated. Choice **(C)** correctly identifies both factors. This eliminates choice **(A)**, which only identifies the first factor. The error in Factor **III** has the opposite effect, eliminating choices **(B)** and **(D)**. If not all of the acid has been titrated because some is left in the flask, the volume of NaOH used will be *too low*, resulting in a molar mass that is *higher* than the true value.

**LC-6.** A student dissolves 60. g of solid NaOH in enough distilled water to make 300. mL of a stock solution. What volume of this solution and distilled water, when mixed, will result in a solution that is approximately 1 M NaOH?

| | Molar Mass, *M* |
|---|---|
| NaOH | 40. g·mol$^{-1}$ |

| | mL stock solution | mL distilled water |
|---|---|---|
| (A) | 20 | 80 |
| (B) | 20 | 100 |
| (C) | 60 | 30 |
| (D) | 60 | 90 |

*Required Knowledge:* (1) Solution vocabulary. (2) Procedures for preparation of solutions in the laboratory. (3) Molarity calculations. $\left( M = \dfrac{\text{mol}_{\text{solute}}}{L_{\text{solution}}} = \dfrac{g_{\text{solute}}}{M_{\text{solute}} \times L_{\text{solution}}} \right)$

*Thinking it Through:* The stock solution contains 1.5 mol of NaOH (60. g divided by the given molar mass of 40 g·mol$^{-1}$) in 300 mL of solution. It is therefore a 5.0 M (1.5 mol divided by 0.300 L) solution. To prepare a solution that is approximately 1 M, a 1:5 dilution will be required. This corresponds to choice **(A)** in which 20 mL of stock solution are mixed with 80 mL of water, forming approximately 100 mL of the diluted solution. Choice **(B)** produces a 1:6 dilution as the total volume is now 120 mL. **(C)** produces a 2:3 dilution and **(D)** produces a 3:5 dilution.

**LC-7.** Nitrogen gas can be collected by displacing mercury from a graduated test tube, as indicated by this apparatus. If the room temperature is 25 °C and the barometric pressure is 760 mmHg, what additional information is most essential in order to calculate the volume of nitrogen under standard conditions?

*Note:* This procedure is no longer commonly performed in the general chemistry laboratory. Using mercury in an open container in this manner creates health and safety concerns. However, the principle illustrated is the same as in any apparatus that uses a column of mercury for measuring the pressure of a trapped gas.

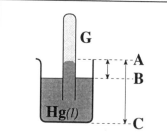

(A)     the height **AC**

(B)     the height **AB**

(C)     the vapor pressure of mercury in region **G**

(D)     the density of nitrogen in region **G**

*Required Knowledge:* (1) Factors affecting the pressure of gases collected by displacing fluids. (2) Molar volume of ideal gas at STP.

*Thinking it Through:* The measured volume of trapped nitrogen must be converted to the volume it would occupy at standard temperature (273 K) and pressure (760mmHg or 1 atm). The level of the mercury in the tube is *higher* than the level of mercury in the beaker. Atmospheric pressure is pushing on the mercury in the open beaker, and the trapped gas is pushing on the mercury remaining in the tube. The relative positions of these two levels indicates that the pressure inside the tube must be *less than* the atmospheric pressure. Since the fluid is mercury, the difference in levels measured in millimeters of mercury can be directly subtracted from atmospheric pressure to determine the pressure of the gas trapped inside the tube. This is correct choice (**B**). It now is possible to convert the volume of this sample of nitrogen to standard conditions of temperature and pressure (273 K and 760 mmHg). Choice (**A**) cannot be the height needed, for it will vary with the size of the container, not with atmospheric pressure. Choices (**C**) and (**D**) give physical properties of the trapped gas, but do not lead to the pressure of the gas that is needed for conversion to standard conditions.

**LC-8.** Three experiments were performed.

(1)  A strip of copper metal was placed in a solution of HCl. No reaction was detected.
(2)  A strip of zinc metal was placed in a solution of HCl. The zinc disappeared and bubbles of gas formed.
(3)  A strip of zinc was placed in a solution of $Mg(NO_3)_2$. No reaction was detected.

Rank the activities of copper, zinc, hydrogen, and magnesium from most active to least active.

(A)     copper > zinc > hydrogen > magnesium

(B)     magnesium > zinc > hydrogen > copper

(C)     zinc > magnesium > hydrogen > copper

(D)     magnesium > copper > hydrogen > zinc

*Required Knowledge:* (1) Evidence for chemical reactions. (2) Interpretation of experimental results to write an activity series.

*Thinking it Through:* Each of the three experiments gives some information about the relative ability of the element to displace an ion in a single-replacement chemical reaction. When a free metallic element reacts with a compound, the free element will replace the metallic ion in the compound, but only if the free element is more reactive than the element it replaces. Here are the two possibilities represented symbolically.

$$\text{If the activity } M_1 > \text{activity } M_2 : M_1 + M_2X \rightarrow M_2 + M_1X$$
$$\text{If the activity } M_1 < \text{activity } M_2 : M_1 + M_2X \rightarrow \text{ no reaction}$$

There was no evidence of chemical reaction in the first experiment. This indicates that hydrogen is a more active element than copper. The second experiment produces bubbles of gas (hydrogen gas) and the zinc strip disappears. This indicates that the zinc replaced hydrogen from the compound, and that zinc is more active than hydrogen. At this point you know that the activity of zinc is greater than that of hydrogen, which in turn is more active than copper. These observations eliminate choices (A) and (D). The third experiment fits magnesium into the series. There is no reaction. Zinc was not successful in replacing magnesium; magnesium is a more active metal than zinc, which is choice (B). This also eliminates choice (C). *Note*: A similar experimental procedure can be used for *nonmetals*, which will replace *nonmetals* in compounds if the free element is more reactive.

---

**LC-9.**   A student combined 1 mL of 0.1 M solutions of nitrate salts of barium, calcium, magnesium, and strontium with several different testing reagents. The observations are recorded in this table. (ppt = precipitate, rxn = reaction.)

| Reagent | $Ba(NO_3)_2$ | $Ca(NO_3)_2$ | $Mg(NO_3)_2$ | $Sr(NO_3)_2$ |
|---|---|---|---|---|
| $H_2SO_4$ | white ppt | no rxn | no rxn | white ppt |
| $Na_2CO_3$ | white ppt | white ppt | white ppt | white ppt |
| $(NH_4)_2C_2O_4$ | white ppt | white ppt | no rxn | white ppt |
| $K_2CrO_4$ | yellow ppt | no rxn | no rxn | no rxn |

A solution contains equal concentrations of barium, calcium, magnesium, and strontium ions. Using the same reagents as in the tests performed, which procedure would best separate the ions by precipitation, followed by filtration?

(A)   Add $K_2CrO_4$, then $H_2SO_4$, then $(NH_4)_2C_2O_4$, then $Na_2CO_3$

(B)   Add $H_2SO_4$, then $K_2CrO_4$, then $Na_2CO_3$, then $(NH_4)_2C_2O_4$

(C)   Add $Na_2CO_3$, then $H_2SO_4$, then $(NH_4)_2C_2O_4$, then $K_2CrO_4$

(D)   Add $K_2CrO_4$, then $(NH_4)_2C_2O_4$, then $Na_2CO_3$, then $H_2SO_4$

---

*Required Knowledge:* (1) Factors governing precipitation formation. (2) Filtering processes. (3) Interpretation of experimental results.

*Thinking it Through:* Ions can often be separated by a sequence of precipitation reactions. The goal is to have only one metal ion precipitate at a time, so that it can be separated by filtration. Then the filtrate be treated with the next reagent. The data indicates that barium ion will be the only ion to precipitate with $K_2CrO_4$, so it is a logical place to start, eliminating choices (B) and (C). The next reagent to precipitate only one ion is $H_2SO_4$ which, of the ions remaining, will only precipitate strontium ion. You now have eliminated choice (D). To be certain that (A) is correct, check that $(NH_4)_2C_2O_4$ can be the next reagent added; it will precipitate the calcium ion. This makes $Na_2CO_3$ the last reagent added, precipitating the magnesium ion. *Note*: This question only depends on the interpretation of reported data, not on the knowledge of any qualitative analysis separation scheme. This is a *qualitative* problem, so it is also not necessary to know or use equilibrium expressions for $K_{sp}$ which may be required for a *quantitative* solution.

---

**LC-10.**   What should be your response, and what is the reason for it, if you enter a laboratory workspace and see this sign?

(A)   Wear your safety glasses; there are corrosive materials in the area.

(B)   Avoid any sudden movement; explosives are stored in the area.

(C)   Do not store organic materials in this area; there are oxidizers present.

(D)   Do not use any open flames; there are flammable materials in the area.

*Required Knowledge:* (1) Laboratory safety signs. (2) Implications for action.

*Thinking it Through:* Working safely in a laboratory requires attention to written, oral, and graphical warnings. There are several signs that are in common use in general chemistry laboratories. You need to know what implicit dangers they represent, and the appropriate response on your part to harm. This sign represents the presence of flammable material, response **(D)**, so it is appropriate not to use open flames.

## Practice Questions

1. The percentage of water in an unknown hydrated salt is to be determined by weighing a sample of the salt, heating it to drive off the water, cooling to room temperature, and re-weighing. Which procedural mistake would result in determining a percentage of water that is too low?

| Procedural Mistakes | |
|---|---|
| I. | heating the sample in a closed, rather than an open, container |
| II. | re-weighing the sample before it has cooled to room temperature |

(A)     I only                                    (B)     II only

(C)     both I and II                          (D)     neither I nor II

2. When an acid solution is titrated with a standard base solution, separate burets are sometimes used for each solution. Which mistake would necessitate emptying and refilling burets, and starting the titration over?

(A)     overshooting the endpoint

(B)     starting with less acid than called for by the procedure

(C)     adding distilled water to the titration flask after a solution was measured into it from the buret

(D)     allowing drops of distilled water to stay in the burets while filling them

3. To prepare 100.0 mL of a 0.100 M copper(II) chloride solution from 0.500 M copper(II) chloride, pipet into a volumetric flask

(A)     10.0 mL of the 0.500 M solution and then add water until the total volume is 100.0 mL.

(B)     20.0 mL of the 0.500 M solution and then add water until the total volume is 100.0 mL.

(C)     20.0 mL of the 0.500 M solution into 80.0 mL of water.

(D)     40.0 mL of the 0.500 M solution into 60.0 mL of water.

4. What is the correctly reported mass of water based on this data?

| Experimental Data | |
|---|---|
| Mass of beaker and water | 29.62 g |
| Mass of beaker only | 28.3220 g |

(A)     1.3 g          (B)     1.30 g          (C)     1.298 g          (D)     1.2980 g

5.  What is the correct reading for the liquid in this buret?

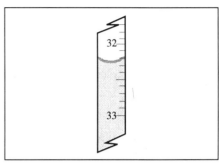

    **(A)**    32 mL      **(B)**    32.2 mL      **(C)**    32.26 mL      **(D)**    33.74 mL

6.  To deliver a 25.00 mL liquid sample most precisely, which piece of glassware would you use?

    **(A)**    volumetric flask                 **(B)**    volumetric pipet

    **(C)**    graduated cylinder           **(D)**    eye dropper

7.  A student determined the percent water in a sample. In four trials, values of 16.145%, 16.160%, 16.156%, and 17.279% were obtained for the percent water in a sample. What value should be used for the reported percent water?

    **(A)**    16.154%      **(B)**    16.435%      **(C)**    16.145%      **(D)**    17.279%

8.  Two students use the same standard mass of 1.0000 g to calibrate the same balance by taking three readings each.

| Reading | Mike's Data | Jean's Data |
|---------|-------------|-------------|
| 1 | 1.0004 g | 0.9996 g |
| 2 | 0.9998 g | 0.9994 g |
| 3 | 0.9992 g | 0.9995 g |

Comparing the two sets of data,

    **(A)**    Mike's is more accurate, but Jean's is more precise.

    **(B)**    Jean's is more accurate, but Mike's is more precise.

    **(C)**    Mike's is both more accurate and more precise.

    **(D)**    Jean's is both more accurate and more precise.

9.  Which procedure can be used to demonstrate experimentally that this reaction obeys the Law of Conservation of Matter?

$$2Mg(s) + O_2(g) \rightarrow 2MgO(s)$$

    **(A)**    Take a mass of 1.0000 g of Mg ribbon, burn it in pure $O_2$, and compare the mass of the product with the original mass of the Mg.

    **(B)**    Show that the sum of two atomic molar masses of Mg plus one molecular molar mass of $O_2$ is equal to two formula molar masses of MgO.

    **(C)**    Determine the mass of a sealed flash-bulb containing magnesium and oxygen, ignite the mixture, cool, and compare the final mass of bulb plus contents with the original mass of the bulb plus contents.

    **(D)**    Burn 1.0000 g of Mg ribbon in a tall beaker filled with air, scrape out all of the MgO formed, and compare with the original mass of the Mg.

**10.** In which case could the given gas be identified by the reaction given?

|  | **Gas** | **Reaction** |
|---|---|---|
| **(A)** | chlorine | will turn starch blue |
| **(B)** | ammonia | will turn moist red litmus paper blue |
| **(C)** | nitrogen | will support combustion |
| **(D)** | hydrogen chloride | will etch glass |

**11.** Gas **A** enters the tube, passes through the contents of **B**, and emerges as gas **C**. If gas **A** is a mixture of carbon monoxide and carbon dioxide, and gas **C** is carbon monoxide, what is most likely substance in the tube?

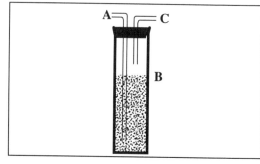

| **(A)** | concentrated HCl(*aq*) | **(B)** | pure water |
|---|---|---|---|
| **(C)** | phosphorous pentoxide | **(D)** | sodium hydroxide solution |

**12.** A 1.0 M solution of an acid is used to titrate several different samples of a 1.0 M base solution. The graph shows the volume of acid needed to neutralize a given volume of base. Which reaction is consistent with the data?

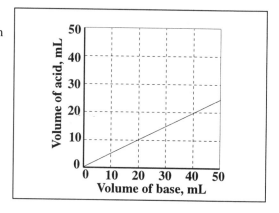

**(A)** $HCl + KOH \rightarrow KCl + H_2O$

**(B)** $2HCl + Ba(OH)_2 \rightarrow BaCl_2 + 2H_2O$

**(C)** $H_2SO_4 + 2KOH \rightarrow K_2SO_4 + 2H_2O$

**(D)** $H_2SO_4 + Ba(OH)_2 \rightarrow BaSO_4 + 2H_2O$

**13.** Which experimental evidence clearly illustrates the oxidizing property of chlorine, $Cl_2$?

**(A)** its liberation at the anode during the electrolysis of sodium chloride solution

**(B)** the reaction of chlorine with sodium bromide solution, resulting in the displacement of bromine

**(C)** the use of chlorates as oxidizing agents in the manufacture of matches

**(D)** its moderate solubility in aqueous solution

**14.** Consider this table of experimental data. In each case, two aqueous solutions were thoroughly mixed, and a small amount of organic solvent was then added, forming a separate layer in the reaction mixture.

| Solution A | Solution B | Color of Organic Layer |
|------------|------------|------------------------|
| NaI(*aq*) | Cl$_2$(*aq*) | purple |
| NaI(*aq*) | Br$_2$(*aq*) | purple |
| NaBr(*aq*) | I$_2$(*aq*) | purple |

What conclusion can be drawn *from these observations*?

**(A)** Iodine is able to displace bromide ion, but not chloride ion.

**(B)** Chlorine is able to displace both bromide and iodide ion.

**(C)** Both chlorine and bromine are able to displace iodide ion.

**(D)** Neither chlorine and bromine are able to displace iodide ion.

**15.** The best method of transferring a coarsely-powdered solid to a six–inch test tube is to

**(A)** pour it through a thin-stemmed glass funnel.

**(B)** pour it from the lip of an evaporating dish.

**(C)** pour it from the bottle originally containing the solid.

**(D)** pour it from a creased square of paper.

**16.** A student combined 1 mL of 0.1 M solutions of nitrate salts of barium, calcium, magnesium, and strontium with several different testing reagents. The observations are recorded in this table.
*Note:* ppt = precipitate, rxn = reaction

| Reagent | Ba(NO$_3$)$_2$ | Ca(NO$_3$)$_2$ | Mg(NO$_3$)$_2$ | Sr(NO$_3$)$_2$ |
|---------|-----------|-----------|-----------|-----------|
| H$_2$SO$_4$ | white ppt | no rxn | no rxn | white ppt |
| Na$_2$CO$_3$ | white ppt | white ppt | white ppt | white ppt |
| (NH$_4$)$_2$C$_2$O$_4$ | white ppt | white ppt | no rxn | white ppt |
| K$_2$CrO$_4$ | yellow ppt | no rxn | no rxn | no rxn |

An unknown solution contains *two* of these four ions: barium, calcium, magnesium, and strontium. The unknown solution forms a white precipitate with H$_2$SO$_4$ and it does not react with K$_2$CrO$_4$. What conclusion about the two ions in the unknown can be drawn from this evidence?

**(A)** It contains only strontium and barium ions.

**(B)** It contains only barium and calcium ions.

**(C)** It contains barium ion and *either* magnesium or calcium ions.

**(D)** It contains strontium ion and *either* magnesium or calcium ions.

**17.** You have a beaker nearly full of gasoline on your laboratory table and it catches on fire. What should you do?

**(A)** Pour water into the beaker.

**(B)** Throw it into the sink.

**(C)** Cover the beaker with a wet cloth.

**(D)** Blow out the flames.

**18.** Which of these pieces of apparatus would be best to use if you want to safely pick up a beaker containing hot liquids?

(A)

(B)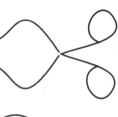

(C)

(D)

**19.** When diluting concentrated $H_2SO_4$, the acid should be added to water because

(A)    concentrated sulfuric acid is a strong acid and can only be added to water.

(B)    concentrated sulfuric acid is such a good oxidizing agent that it is able to oxidize water.

(C)    concentrated sulfuric acid must be protected from air by a layer of water.

(D)    sconcentrated sulfuric acid and water generates considerable heat when in contact.

**20.** What should your response be if you enter an unfamiliar laboratory workspace and see this sign?

(A)    You should immediately evacuate the premises and then call for OSHA.

(B)    You should put on full protective clothing, including goggles, respirator, and body suit.

(C)    You should be aware that chemical waste is being processed in the immediate area.

(D)    You should pay careful attention to your surroundings for potential hazards.

## Answers to Study Questions

| | | | | | |
|---|---|---|---|---|---|
| 1. | A | 5. | C | 9. | A |
| 2. | D | 6. | A | 10. | D |
| 3. | D | 7. | B | | |
| 4. | D | 8. | B | | |

## Answers to Practice Questions

| | | | | | | | |
|---|---|---|---|---|---|---|---|
| 1. | A | 6. | B | 11. | D | 16. | D |
| 2. | D | 7. | A | 12. | C | 17. | C |
| 3. | B | 8. | A | 13. | B | 18. | B |
| 4. | B | 9. | C | 14. | C | 19. | D |
| 5. | C | 10. | B | 15. | D | 20. | D |

# GENERAL CHEMISTRY DATA SHEET
## ACS Examinations Institute

| ABBREVIATIONS AND SYMBOLS | | | | | |
|---|---|---|---|---|---|
| amount of substance | $n$ | equilibrium constant | $K$ | molal | $m$ |
| ampere | A | Faraday constant | $F$ | molar | M |
| atmosphere | atm | free energy | $G$ | molar mass | $M$ |
| atomic mass unit | u | frequency | $\nu$ | mole | mol |
| atomic molar mass | $A$ | gas constant | $R$ | Planck's constant | $h$ |
| Avogadro constant | $N_A$ | gram | g | pressure | $P$ |
| Celsius temperature | °C | hour | h | rate constant | $k$ |
| centi– prefix | c | joule | J | second | s |
| coulomb | C | kelvin | K | speed of light | $c$ |
| electromotive force | $E$ | kilo– prefix | k | temperature, K | $T$ |
| energy of activation | $E_a$ | liter | L | time | $t$ |
| enthalpy | $H$ | measure of pressure | mmHg | volt | V |
| entropy | $S$ | milli– prefix | m | volume | $V$ |

| CONSTANTS |
|---|
| $R = 8.315 \text{ J·mol}^{-1}\text{·K}^{-1}$ |
| $R = 0.0821 \text{ L·atm·mol}^{-1}\text{·K}^{-1}$ |
| $1\ F = 96,500 \text{ C·mol}^{-1}$ |
| $1\ F = 96,500 \text{ J·V}^{-1}\text{·mol}^{-1}$ |
| $N_A = 6.022 \times 10^{23} \text{ mol}^{-1}$ |
| $h = 6.626 \times 10^{-34} \text{ J·s}$ |
| $c = 2.998 \times 10^8 \text{ m·s}^{-1}$ |
| $0\ °C = 273.15 \text{ K}$ |

# PERIODIC TABLE OF THE ELEMENTS

| 1 1A | 2 2A | 3 3B | 4 4B | 5 5B | 6 6B | 7 7B | 8 8B | 9 8B | 10 8B | 11 1B | 12 2B | 13 3A | 14 4A | 15 5A | 16 6A | 17 7A | 18 8A |
|---|---|---|---|---|---|---|---|---|---|---|---|---|---|---|---|---|---|
| 1 H 1.008 | | | | | | | | | | | | | | | | | 2 He 4.003 |
| 3 Li 6.941 | 4 Be 9.012 | | | | | | | | | | | 5 B 10.81 | 6 C 12.01 | 7 N 14.01 | 8 O 16.00 | 9 F 19.00 | 10 Ne 20.18 |
| 11 Na 22.99 | 12 Mg 24.31 | | | | | | | | | | | 13 Al 26.98 | 14 Si 28.09 | 15 P 30.97 | 16 S 32.07 | 17 Cl 35.45 | 18 Ar 39.95 |
| 19 K 39.10 | 20 Ca 40.08 | 21 Sc 44.96 | 22 Ti 47.88 | 23 V 50.94 | 24 Cr 52.00 | 25 Mn 54.94 | 26 Fe 55.85 | 27 Co 58.93 | 28 Ni 58.69 | 29 Cu 63.55 | 30 Zn 65.39 | 31 Ga 69.72 | 32 Ge 72.61 | 33 As 74.92 | 34 Se 78.96 | 35 Br 79.90 | 36 Kr 83.80 |
| 37 Rb 85.47 | 38 Sr 87.62 | 39 Y 88.91 | 40 Zr 91.22 | 41 Nb 92.91 | 42 Mo 95.94 | 43 Tc (98) | 44 Ru 101.1 | 45 Rh 102.9 | 46 Pd 106.4 | 47 Ag 107.9 | 48 Cd 112.4 | 49 In 114.8 | 50 Sn 118.7 | 51 Sb 121.8 | 52 Te 127.6 | 53 I 126.9 | 54 Xe 131.3 |
| 55 Cs 132.9 | 56 Ba 137.3 | 57 La 138.9 | 72 Hf 178.5 | 73 Ta 180.9 | 74 W 183.8 | 75 Re 186.2 | 76 Os 190.2 | 77 Ir 192.2 | 78 Pt 195.1 | 79 Au 197.0 | 80 Hg 200.6 | 81 Tl 204.4 | 82 Pb 207.2 | 83 Bi 209.0 | 84 Po (209) | 85 At (210) | 86 Rn (222) |
| 87 Fr (223) | 88 Ra (226) | 89 Ac (227) | 104 Rf (261) | 105 Db (262) | 106 Sg (263) | 107 Bh (262) | 108 Hs (265) | 109 Mt (266) | 110 (269) | 111 (272) | | | | | | | |

| 58 Ce 140.1 | 59 Pr 140.9 | 60 Nd 144.2 | 61 Pm (145) | 62 Sm 150.4 | 63 Eu 152.0 | 64 Gd 157.3 | 65 Tb 158.9 | 66 Dy 162.5 | 67 Ho 164.9 | 68 Er 167.3 | 69 Tm 168.9 | 70 Yb 173.0 | 71 Lu 175.0 |
|---|---|---|---|---|---|---|---|---|---|---|---|---|---|
| 90 Th 232.0 | 91 Pa 231.0 | 92 U 238.0 | 93 Np (237) | 94 Pu (244) | 95 Am (243) | 96 Cm (247) | 97 Bk (247) | 98 Cf (251) | 99 Es (252) | 100 Fm (257) | 101 Md (258) | 102 No (259) | 103 Lr (262) |